T0248224

DR. CALHOUN'S MOUSERY

DR. CALHOUN'S MOUSERY

THE STRANGE TALE OF A CELEBRATED SCIENTIST, A RODENT DYSTOPIA, AND THE FUTURE OF HUMANITY

LEE ALAN DUGATKIN

The University of Chicago Press CHICAGO AND LONDON

The University of Chicago Press, Chicago 60637
The University of Chicago Press, Ltd., London

Published 2024
Printed in the United States of America

33 32 31 30 29 28 27 26 25 24   1 2 3 4 5

ISBN-13: 978-0-226-82785-8 (cloth)
ISBN-13: 978-0-226-82786-5 (e-book)
DOI: https://doi.org/10.7208/chicago/9780226827865.001.0001

Library of Congress Cataloging-in-Publication Data

Names: Dugatkin, Lee Alan, 1962– author.
Title: Dr. Calhoun's mousery : the strange tale of a celebrated scientist, a rodent dystopia, and the future of humanity / Lee Alan Dugatkin.
Description: Chicago : The University of Chicago Press, 2024. | Includes bibliographical references and index.
Identifiers: LCCN 2024011409 | ISBN 9780226827858 (cloth) | ISBN 9780226827865 (ebook)
Subjects: LCSH: Calhoun, John B. | Mice—Behavior. | Overpopulation.
Classification: LCC QL752 .D84 2024 | DDC 599.35—dc23/eng/20240405
LC record available at https://lccn.loc.gov/2024011409

♾ This paper meets the requirements of ANSI/NISO Z39.48-1992 (Permanence of Paper).

FOR LENA AND JULIA RENE I

But Mousie, thou art no thy-lane,
In proving foresight may be vain:
The best laid schemes o' Mice an' Men
Gang aft agley,
An' lea'e us nought but grief an' pain,
For promis'd joy!

ROBERT BURNS, "To a Mouse, on Turning Her Up in Her Nest, with the Plough," November 1785

CONTENTS

PREFACE

As the author of an animal behavior textbook and coauthor of a textbook on evolutionary biology, I've read thousands of articles on behavior, population biology, and evolution, many published in the very best journals science has to offer. None of those papers begins with a more unusual, arguably—or perhaps not so arguably—bizarre, opening sentence than that found in John B. Calhoun's 1973 paper, "Death Squared: The Explosive Growth and Demise of a Mouse Population." "I shall largely speak of mice," Calhoun began that paper in the *Proceedings of the Royal Society of Medicine*, "but my thoughts are on man, on healing, on life and its evolution." Even more remarkably, these are the same words Calhoun uttered a year earlier, when he lectured on his "Death Squared" paper at the Royal Society of Medicine in London.

I can't recall for certain when I first encountered John Calhoun's work, but the odds are good that it was when I was in graduate school in the late 1980s. My thesis was on cooperation in nonhumans, and I read every related thing I could get my hands on, including E. O. Wilson's 1975 behemoth of a book, *Sociobiology*, which I devoured cover to cover. In *Sociobiology*, Wilson discusses Calhoun's work using, at least from a graduate student's perspective, a pair of tantalizing adjectives—"famous" and "bizarre"—the former when describing

"Calhoun's . . . Norway rat colonies," and the latter when discussing the "effects [that] were observed by Calhoun."

If that was indeed my first exposure to Calhoun's studies, it resulted in no more than my storing that work somewhere in the back of my mind until 2001, when I began writing the first edition of my textbook, *Principles of Animal Behavior* (now in its fifth edition). One of many things I wanted to do with *Principles of Animal Behavior* was give students a better perspective on the history of the science of animal behavior than was available in other books at the time, and to do that I read or reread lots of material from the early years of the study of animal behavior. My memory was jolted, as again I came across a few references to Calhoun's work. Tempted though I was, with all I needed to do write the first edition of a textbook, I felt I didn't have time to dive deeper and learn more about that work. In retrospect, I wish I had made the time.

Fast forward close to another two decades to 2020, and my memory was jolted again. This time I was contemplating writing a book about social networks in nonhumans when I came across a series of papers about social networks in mice. That work, like Calhoun's early study on rats, took place in a barn (this time near Zurich, Switzerland). And those studies on social networks weren't only about complex social dynamics but population growth as well. Just like Calhoun's work had been. This time I acted and did a dive into Calhoun and his experiments. That began by reading a handful of Calhoun's papers, then more. Soon, I was gathering and scanning scores of newspaper articles written about Calhoun's work while it unfolded and taking notes based on the fine works that historians Edmund Ramsden and Jon Adams had written about Calhoun more recently.

I was more than a bit intrigued by what I was learning. I came to know that Calhoun was taking lessons from work on population growth and population crashes in mice and rats and applying them to overpopulation in our own species—warning that we would cause our own extinction if we didn't act. And soon. "Rats and mice, of course, are not perfect models for humans," Calhoun told a *Washington Post* reporter, "but the disaster they represent is so compelling that the world cannot wait for proof of every step in the equation." As if that was not interesting enough, I was digging up bits and pieces intimating that Calhoun's studies on rodents and population growth had, of all things, led him to propose an early version of the world wide web. After next sifting through hundreds of items in the John B. Calhoun archives at the National Library of Medicine in Bethesda, Maryland—items that included Calhoun's grant proposals, unpublished manuscripts, and personal correspondence—I knew there was no turning back. I was hooked and convinced that this story needed to be told, and the sooner the better.[1]

INTRODUCTION

Standing before the Royal Society of Medicine in London on June 22, 1972, ecologist-turned-psychologist John Bumpass Calhoun, director of the Laboratory of Brain Evolution and Behavior at the National Institute of Mental Health, appeared a mild-mannered, smallish man, sporting a graying goatee. After what must surely have been one of the oddest opening remarks to the Royal Society in its storied two-hundred-plus-year history, Calhoun spoke of a long-term experiment he was running on the effects of overcrowding and population crashes in mice.

Members of the Royal Society were scratching their heads as Calhoun told them of Universe 25, a giant experimental setup he had built and which he described as "a Utopian environment constructed for mice." Still, they listened carefully as he described that universe. They learned that to study the effects of overpopulation, Calhoun, in addition to being a scientist, needed to be a rodent city planner. For Universe 25, he had built a large, very intricate apartment block for mice. There were sixteen identical apartment buildings arranged in a square with four buildings per side. Calhoun told his audience each building had "four four-unit walk-up one-room apartments," for a total of 256 units, each of which could comfortably accommodate about fifteen mouse residents. There were also a series of dining halls in each apartment building, and a cluster of rooftop fountains to quench the residents' thirst. Calhoun had marked each mouse resident with a unique color combination and he or his team sat in a loft over this mouseopolis, for hours every day for more than three years, and watched what unfolded.

Calhoun told the Royal Society members that what began as a rodent utopia—where mice had sumptuous accommodations, all the food and water they could want, and were free from the twin scourges of disease and predation—in time degenerated into a mouse hell. Initiated by a population explosion early on, and later stagnation and decline, that hell had mice displaying a suite of aberrant behaviors, including the loss of sexual drive on the part of males and the absence of maternal care in females. Calhoun attributed much of this to the formation of what he called a *behavioral sink*: a "pathological togetherness" that developed among the mice in Universe 25. "Normal social organization . . . 'the establishment,'" he told the crowd, "breaks down, it 'dies.'"

Even if mice were taken from Universe 25 and placed into another mouse apartment block with a much lower density of residents, Calhoun explained, they still showed these aberrant behaviors. As he summarized his results, Calhoun again befuddled his audience, telling them of a class of mice that had appeared after the rodent population bomb had exploded. These were what Calhoun dubbed "the beautiful ones," who spent their time grooming themselves and eating and shunned all social behavior. The Beautiful Ones, Calhoun told his audience, were "capable only of the most simple behaviours compatible with physiological survival."

Calhoun's work a decade earlier, that time on rats in a barn-turned-laboratory, had already been the subject of much attention, garnering stories in major newspapers around the world—but now, in conjunction with the mice in Universe 25, his studies of crowding, population growth, and the perils of overpopulation in rodents skyrocketed him to international attention.

No one in attendance at Calhoun's 1972 "Death Squared" lecture at the Royal Society of Medicine would soon forget what they heard that day. Some of the society members, certainly those who had invited him, knew of Calhoun's article in *Scientific American* a decade earlier. That article, on crowding and population crashes in rats, opened with allusions to eighteenth-century political economist Thomas Malthus and his ideas on overpopulation and misery. Calhoun then followed with a description of his own results in rats, which centered on the effects of overcrowding and unconstrained population growth. But the Universe 25 results he told the Royal Society of that June day in 1972 were different. And not just because of the Beautiful Ones, or because these results confirmed what Calhoun found in the rats a decade earlier, but because the work was so much grander, with thousands of mice and many years of data on a topic—population growth and decay—that was of great concern, and not just to scientists. A large swath of the public at the time was terrified of the implications of human population growth as well—largely because of Paul Ehrlich's 1968 book, *The Population Bomb*, which shot to the bestseller list after Ehrlich appeared on *The Tonight Show* with Johnny Carson.[1]

Calhoun's work was covered in the *New York Times*, the *Washington Post*, *Time* magazine, *Der Spiegel*, and more: "It was a lovely day," read the opening sentence of a 1970 *Newsweek* story, "much too lovely to spend in an office. In fact, it seemed the perfect day to visit Dr. John Calhoun's mousery." In *Time* magazine's 1971 story, "Population Explosion: Is Man Really Doomed?," the writer notes grimly that "even if some way can be found to feed the onrushing millions [of humans], they may still face a psychic fate similar to the one that befell Dr. John Calhoun's white mice." In April of that year, the United States Senate discussed Calhoun's works on overpopulation in rodents and the implications for our own species, and it enshrined three of his papers in the *Congressional Record*.

Many saw Calhoun's work on overcrowding and population dynamics as a portent of doomsday, but he thought otherwise. Calhoun came to see the results of his work with rodents as prescriptive, showing a path forward to prevent the human population bomb from exploding. Many agreed, but he had his critics, then and now. He dubbed himself a ℞evolutionist (note the first letter) and his ideas as "metascientific," which, for Calhoun, meant using science to try to understand very complex problems involving not just many unknowns but complex interactions between variables. To drive home that complexity, he proposed experiments in rats that would tinker with rodent culture and cooperation, and in so doing, defuse a potential ticking population bomb.

Every edition of *Forty Studies That Changed Psychology*, from the first in 1992 to the latest in 2020, has a chapter devoted to Calhoun's work. But, in many ways, the most profound impact of Calhoun's studies lies far from academic halls and ivory towers. Through the seemingly endless coverage in newspapers and magazines, Calhoun's work seeped into the public consciousness. On the practical end, city planners and architects in the 1970s and 1980s looked to Calhoun's results when designing housing developments, and he encouraged them to do so. Each year, Ian McHarg, who years later was awarded the National Medal of Arts and the Thomas Jefferson Foundation Medal in Architecture, brought Calhoun to lecture at the Interdisciplinary Program of Landscape Architecture at the University of Pennsylvania. Calhoun hadn't always talked of mice while thinking of men, but by the mid-1970s, results from his studies over the previous twenty-five years led him to believe he had a moral responsibility to do so.

Film producers and both fiction and nonfiction writers clamped onto Calhoun's ideas and incorporated them into their own work. Page after page of Tom Wolfe's 1968 book *The Pump House Gang*—published on the same day his *The Electric Kool-Aid Acid Test* hit the stands—paid homage to Calhoun's idea of a "behavioral sink." Calhoun's work, in part, led one of the writers of *Catwoman* to introduce the character Ratcatcher to the comic strip. The best-selling children's book *Mrs. Frisby and the Rats of NIMH* may have had its origin in Calhoun's rat experiments: at the very least, Calhoun thought it did, based on a visit by the author, Robert Conly (who wrote under the pseudonym Robert C. O'Brien), to his lab at the National Institute of Mental Health (NIMH). And the list goes on and on.

In perhaps the oddest twist of all, as early as the 1960s, Calhoun's studies of mice and rats led him to devote a good deal of the later years of his career to working on the creation of a human "world brain." "We are now at the critical transition to a new type of man," Calhoun told an audience in 1969, "one which depends increasingly on extracortical prostheses to evolve and utilize concepts." Calhoun was convinced that by linking together these extracortical prostheses, we might be able to harness creativity and, among other things, figure a way out of the overpopulation problem. Never one to shy away from bold prediction, Calhoun told this same audience that "a rough calculation indicates that by 40,000 years from now less than 5% of creative activity will be done by our cortex and at least 95% by prostheses."[2]

Over the last fifty years, Calhoun's work has gradually fallen off the map in academic circles, but it continues, to this day, to live on in the public psyche. Why? Who was Calhoun, and what drove him to study crowding and overpopulation in rats and mice, and to use that work as a portal for both warning about the existential perils of human overpopulation and for proposing a world brain to help us avert that peril? What on earth was going on with all this?

To start to piece together the answers we need to step back in place and time, briefly to the eighteenth century and then to rural Tennessee in 1917.

1

LET'S GO WATCH THE BIRDS

In the late 1750s, one band from the Calhoun clan set sail from Scotland. Upon disembarking in North America, they settled in Long Cane Creek, South Carolina. A few years later, a branch of that colonial lot of Calhouns headed to North Carolina, in time also spreading to Tennessee, while another branch remained put. The South Carolinian side of the family produced secessionist firebrand John C. Calhoun. The other, from which John Bumpass Calhoun is descended, "tended to produce teachers and preachers," wrote John B., "to retreat into the world of ideas."

Born on May 11, 1917, in rural Tennessee, in a speck of a town called Elkton, John B. Calhoun was fond of referring to himself as a "Tennessee country boy." Barely two square miles at the time of Calhoun's birth, Elkton's population was about 175, and its claim to fame was that General William Tecumseh Sherman camped there one night on his Civil War march to Atlanta.

John's parents were James and Fern Madole Calhoun: James was a high school principal, and work often had him on the road visiting nearby schools, as was the case when Fern was nine months pregnant with John. "Naturally, the home situation was on his mind," according to Fern, who insisted he go on the business trip, as the family needed the money. "At a store in this neighboring town," she continued, "when the druggist asked him how many children he had, James replied 'I don't know.'" The timing of the question was fortuitous, as soon thereafter a clerk in the store asked if a Mr. Calhoun was there, as there was a call for him. When James put the phone down, "he turned to the druggist and said, 'I know now—I have two.'"

When John Calhoun was a year old, his family moved sixty miles east to Winchester, where his father had taken on the principalship in a larger high school. More important to John's development, Fern opened a small kindergarten in their new house and so served as both mother and teacher to the young boy. It was in kindergarten that John showed the first hints of an interest in the natural world. The other students' requests to "sing and dance," his mother recalled, "were as frequent as [John's] 'let's go watch the birds.'" Fern tried to find something that captured the attention of each individual child in her class, and John's "first memory of an attraction to animals hinged on sitting with her on the ground in a sun flecked woodland looking at a lizard she had just caught and held in her handkerchief." Fern was equally adept at inspiring the others in her class: one of the kindergarteners who preferred singing and dancing was Dinah Shore, who went on to use those skills in a wildly successful career in recording, radio, film, and television, crediting Fern Calhoun with igniting her passion for the performing arts.

Following a new principal position opening for James, when John completed kindergarten, the Calhouns moved again, this time to Brownsboro, Tennessee, population three thousand. John remembered a "physically isolated county, festooned with traditions," where students came to school riding horses, and he had an elementary school teacher who liked nothing more than "chewing on apple seeds and regaling his students of his own boyhood days during the Civil War." John was a sharp student, but a loner who preferred spending time romping around in the woods to playing with the other children. "Hunting—we received shotguns early—and collecting bird eggs," he recalled, "sharpened my powers of observing nature during these elementary years." John's most vivid memories of those years were of "watching doves and quail in flocks . . . following the trails of cottontail rabbits through brush and fences . . . getting up at dawn and watching the light-loving animals become active . . . building bluebird boxes . . . [and] making a collection of butterflies." In his spare time, young John tended to his own garden and sold its produce door to door, using the money he made to buy a sow, whose piglets he turned over for a nice profit.[1]

The last move of young John Calhoun's early life came in 1930, at the height of the Great Depression, when he was thirteen. His father had been promoted to the school board and charged with running a statewide schoolhouse consolidation. The move took the family to big-city Nashville, and it was during his time there at Hume Fogg High School that the young man, who as a kindergartener wanted nothing more than to go out and watch the birds, became a full-fledged junior birder, as well as a budding ecologist. While John's father worked on a master of planning degree at Vanderbilt's Peabody College, the family moved into a dormitory on campus for a short stint. There, John met several biology professors, including Jesse Shaver, who shared the young man's interest in ornithology and started to take John, binoculars in hand, on field trips. With John's passion for animals now bubbling over, family and friends started feeding it, giving him books about birds and mammals, including Roger Torey Peterson's 1934 *Field Guide to Birds*, which he quicky devoured.

Jesse Shaver was impressed enough with the teenaged Calhoun's ability to handle himself in the field that he began taking him to meetings of the Tennessee Ornithological Society, where Calhoun met many like-minded types, including Anna Cochran and Amelia Laske, who were interested in banding and studying chimney swifts (*Chaetura pelagica*), common birds in the region. They taught Calhoun how to trap the birds and band them, and his size proved to be an asset—"being small," his mother recalled, "[John] seemed to have an advantage scaling the chimneys, but [he] let it be known such adventures gave great concern to his parents." Soon enough Calhoun was tinkering, building his own traps, and sampling the birds near his home. He also built a pigeon coop in the loft of a nearby neighbor, cultivated a garden that he seeded with moss and wildflowers that he brought back from birding trips, and constructed a pond in the backyard, stocking it with terrapin turtles.

In 1933, Calhoun's father lost his school board job when the governorship of Tennessee switched hands. This jeopardized John's prospects for attending college. Then serendipity stepped in. Some members of the Tennessee Ornithological Society were driving up to Cape Cod to do a bit of birding, and they told Calhoun that he could hitch a ride with them for as far as he liked. Calhoun's friend, Jack Hayes, was spending the summer at the University of Virginia, where his father, a professor of sociology, was teaching a summer course. Calhoun thought it would be fun to visit, so he took the society members up on their offer and jumped off at Charlottesville.

As Calhoun and his friend walked around the famous lawn of Thomas Jefferson's Academical Village, with its colonnaded pavilions, they bumped into Ivey Lewis, a colleague of Hayes's father and a dean at the university. Lewis, a geneticist and botanist, and one of the founders of the Virginia Academy of Science, was also an avid birder and took an immediate liking to Calhoun. "On the spot, within ten minutes," as Calhoun recalls the meeting, "he offered me a scholarship [to the University of Virginia] and an NYA [National Youth Administration] job." A product of the New Deal, the NYA eventually morphed into the "work study" programs that are ubiquitous at American universities today. Calhoun lined up room and board with his friend Jack's grandparents, who had a farm near the university, and between the dean's scholarship, the NYA support, and selling some parts of an egg collection he had amassed in high school, Calhoun had what he needed to enroll as a full-time student.

Calhoun immediately plunged into the study of biology and was awestruck by the things he was learning. In his freshman biology course, his instructor told students of H. V. Wilson's work on the regenerative ability of sponges. In a 1907 paper, Wilson described a study with the red oyster sponge (*Microciona prolifera*), in which he first cut a sponge into pieces a quarter of an inch in diameter and then pushed the pieces through a fine mesh, breaking each bit of sponge into a myriad of single cells and creating "the appearance of red clouds" in the water. Cells began to join back together almost immediately thereafter, and clusters of cells sat where their constituent parts had sat just a few minutes prior. Calhoun sat and listened to this story of annihilation and rebirth and contemplated what that experiment said about animal and human societies being broken into pieces, and what forces, if any, led to their reconstitution.

Two other University of Virginia biologists—Alfred Kepner and Orland White—also left their mark on the young Calhoun: "From Kepner came the conviction that one can detect a design, and perhaps even purpose, in life," he wrote, "and from White the knowledge that out of randomness, organization and evolution can emerge." In time, Calhoun became a fixture at the "all-day breakfasts" White would host for his graduate students, where "sharpening one's logic in the wide-ranging discussions" was the order of the day.

More than just for paying tuition, Calhoun was able to use his NYA work-study position in pursuit of a scientific education. To earn that money, he was tasked with obtaining bird and small-mammal carcasses and collecting the parasites within to be used in general biology and the parasitology courses. This gave him much-needed time in the woods setting traps for mice and shrews, and on occasion shooting birds, but it also created contacts in the Biology Department that allowed him to sit in on graduate seminars. In one seminar, he heard a graduate student review Alfred Kinsey's new book all about gall wasps. Long before his work on human sexology, Kinsey was a world-renowned entomologist, and after the seminar that day, Calhoun read Kinsey's wasp book from cover to cover.

Though Calhoun fit right in academically, it took time for the self-described Tennessee country boy to learn how to navigate the culture of his new surroundings at the university. His lone attempt to engage in student politics did not go well. In the mid-1930s, the American Youth Congress was speaking up for the rights of the country's young people and their work, which, in part, led Eleanor Roosevelt to play a role in creating the NYA that was helping fund Calhoun. A spin-off organization from the American Youth Congress was the American Student Union, which had a branch at the University of Virginia. Apparently unaware of its ultra-leftist philosophy, Calhoun joined but quickly resigned when he discovered that his politics did not align with theirs, as all members were avowed communists.

It was more than just a naive sortie into the world of college politics that posed problems. The university was "a 'gentleman's' school," Calhoun wrote. "The honor system, coats and ties . . . neither by budget nor temperament did this formal attire suit me. . . . As a consequence, I was occasionally threatened with physical violence . . . if I didn't conform." Calhoun wasn't intimidated. The fact that he was sometimes seen walking around campus with the 16-gauge shotgun he used when collecting birds and small mammals (to gather parasites for the biology and parasitology courses) seemed to convince his fellow students that he was someone that they might be better off not harassing.[2]

Calhoun's escape continued to be birding. He kept his connections with the Tennessee Ornithological Society, in particular with Amelia Laske, who had initiated work on chimney swifts back when Calhoun had been in Nashville. In the summer of 1936, not yet out of his teens, sophomore Calhoun, working with Laske and others, used his summer vacation to undertake a large-scale project banding the swifts in Nashville, in part to study the birds' movements.

Using a large trap that he and a friend had constructed to make it difficult for birds to get out once they were inside, Calhoun set to work that included climbing chimneys, like that of the Blakemore Methodist Church, home to about eight hundred swifts. Calhoun would light a fire in a smudge pot and lower it down the chimney, forcing the birds to exit. For smaller chimneys, he'd lower down a pot full of rocks and rattle it about until the birds departed. When the swifts returned, the trap, which could hold many a bird, was in place at the opening of the chimney; it was eventually lowered to the ground, where the birds were banded with unique leg tags. Developing an eye for detail, Calhoun noted that birds entering a chimney did so in one of two very different ways: either hundreds of swifts would circle clockwise or counterclockwise and then, in a flash, all enter the chimney virtually simultaneously or they entered one at a time, rather than en masse.

Not much was known of the movement patterns of swifts, so Calhoun repeated his capture and banding process over and over, both in Nashville and in Clarksville, about fifty miles northwest. In this, his first lesson about animal population dynamics, Calhoun discovered that flock membership was fluid. Swifts would search for food during the day and then roost in a nearby chimney with whoever else happened to be roosting there that evening: the *number* of birds in a chimney stayed relatively constant over time, but *who* was at the chimney did not.

Foraging sorties would often take the swifts far from where they started out in the morning, and Calhoun recorded dozens of marked birds that were captured at sites in both Nashville and Clarksville. In 1938, during his junior year at the University of Virginia, Calhoun published his first scientific paper, a four-page article titled "Swift Banding at Nashville and Clarksville." A second paper followed in which, working with ornithologists all along the East Coast, Calhoun and his colleagues mapped the seasonal migrations of chimney swifts.

By the time he was ready to graduate, Calhoun had as much experience in field ecology as many professionals in that nascent discipline. The question was, What was next? He had thought about graduate school in ecology, but not where to do that work, when one evening he bumped into a fellow ecology student, J. C. Strikland. Strikland told Calhoun he was thinking of doing his own graduate work at the Field Museum in Chicago and then mentioned that he had recently thrown away a flyer from the chairman of the Zoology Department at Northwestern University, a short fifteen miles north of the Field Museum, advertising a teaching assistant position. Calhoun prodded Strikland into retrieving the letter from the trash. Despite not having the master's degree that the advertisement called for, Calhoun applied, hoping that his published papers, field experience, and time sitting in on graduate seminars would stand in place of a master's. Northwestern liked what they read in the young ecologist's resume and offered him the position and a spot in their PhD program.

Before starting at Northwestern, Calhoun took the summer immediately following graduation to go back out into the woods, this time as part of a Tennessee Department of Conservation team charged with surveying the birds and mammals of the Hatchie River drainage in the southwest region of the state. One paper that came from that survey has Calhoun as sole author, and lists him as "John Calhoun, ornithologist." In that paper, Calhoun recorded more than a hundred species of birds that he encountered in the drainage, but equally important, in many instances he provided information on their ecology, including feeding behavior, activity patterns, estimates of population size, and more. The team that summer along the Hatchie River included camp managers, a cartographer, and mammologist Willet Wandell, who did a lot of work capturing and cataloging small mammals, like those Calhoun had collected for classes at the University of Virginia. No doubt, sitting around the campfire that summer of 1939, Wandell led many a late-night conversation about the mice and voles he encountered that day. Those conversations would serve Calhoun well in graduate school and beyond.[3]

2

UNDER TUTELAGE

In 1939, as Calhoun was finishing his time at the University of Virginia, ecology was a very young discipline. On occasion, ecology papers could be found in broad-based journals like *Nature* or *Science*, and more often in the *American Naturalist*, but the British Ecological Society and the Ecological Society of America, and their respective journals, were still relative newcomers. The very term ecology—derived from the Greek word *oikos* ("home" or "house")—had only been coined by Ernst Haeckel in the late 1860s, and early European and American ecologists interpreted Haeckel's discipline as the study of the "general economy of the household of nature," "scientific natural history," or the "the sociology of organisms."

The same year Calhoun began his graduate work in ecology at Northwestern University, McGraw-Hill published the second edition of Arthur Sperry Pearse's textbook, *Animal Ecology*. At the opening of the book, Pearse warned readers there were still ecologists clinging to a vitalist view of nature and arguing that "an animal consists of matter, energy, and also some vital principle that cannot be weighed or measured." Though not completely dismissive of the possibility, Pearse reassured his readers that "most scientists, whatever their beliefs, proceed as mechanists in that they attempt to find scientific explanations for all phenomena associated with life." But as a young graduate student in ecology who already had a penchant in that direction from his work with swifts, Calhoun would have been struck by Pearse's emphasis on population-level thinking. After discussing how food, shelter, predators, temperature, and so on affect population growth, Pearse presented the widely held view that animal (and plant) population size tends to skyrocket when a species first enters a new environment then continues to increase at a more tempered pace before finally settling down to a size where, all else being equal, it remains. The idea that populations behave this way, displaying what's called logistic growth, was, in large part, due to Raymond Pearl, an animal behaviorist turned statistical demographer.[1]

After completing his dissertation work on the behavior of aquatic worms in 1902, Raymond Pearl spent a year studying with Karl Pearson, a leader in the new field of biometry—the statistical study of biological phenomena. Soon Pearl became a biometry wunderkind and in 1918 was recruited as the first professor of biometry at Johns Hopkins University's newly minted School of Hygiene and Public Health. There, he focused on the study of longevity, breeding fruit flies that he obtained from Thomas Hunt Morgan's already famous "fly room" at Columbia University. Two years after arriving at Johns Hopkins, Pearl, together with Lowell Reed, published a paper in the *Proceedings of the National Academy of Sciences* in which they laid out a model of logistic population growth. Pearl describes what this type of growth looks like on a graph where population size is tracked through time. Imagine, Pearl tells his readers, that you have a flexible wire cutout of the letter *S*. Stand the *S* up on a flat surface, press down and stretch the top to the right and the bottom to the left, and you get a shape that resembles logistic growth: rapid growth early, followed by slower growth, finally ascending to a population plateau that is called the "carrying capacity." The carrying capacity of a population is determined by the particular environment in which individuals in that population live; as such, it depends on the amount

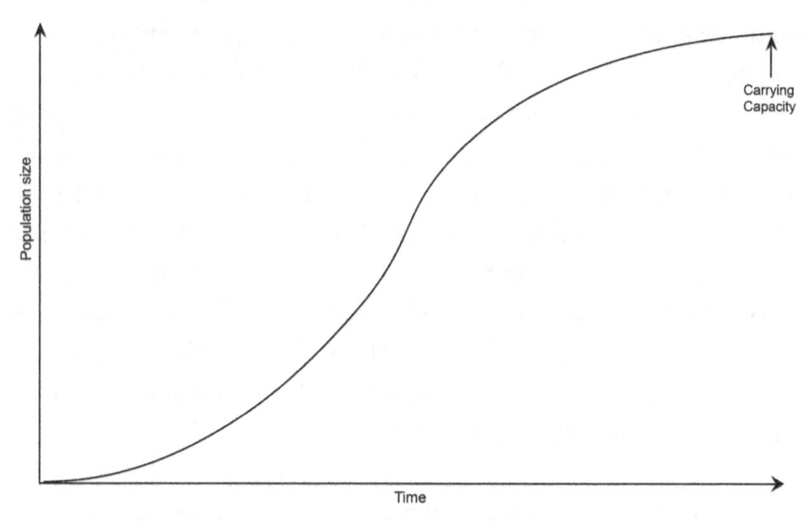

Logistic growth. A population grows exponentially early on, after which growth slows and eventually plateaus at the population's carrying capacity.

and distribution of food and water, the risk of predation and disease, and more. Below the carrying capacity, populations are predicted to increase and above it, to decrease.

Much of the data in the *Proceedings of the National Academy of Sciences* paper came from human population growth rates in the United States, but soon Pearl suggested the general shape of the logistic curve applied to population growth across the wider animal kingdom, as well as to plants and microbes. No wonder, then, that when students like John Calhoun read the chapters on population growth in their ecology textbooks, what they learned about was logistic growth. Those same chapters also taught students that it was only when population size overshoots the carrying capacity that trouble sets in. Overpopulation, and particularly crowding, might lead to a population gradually decreasing back to capacity, but it could also lead to the spread of disease, the killing of young, and more—potentially resulting in a population plummeting far below carrying capacity, perhaps down to extinction. With one important exception, which we will return to shortly, the idea that underpopulation could have adverse effects was not on the radar of ecologists.[2]

Forsaking the Tennessee-country-boy-with-a-shotgun-in-hand look that served him well at the University of Virginia, Calhoun, now a graduate student with teaching assistant responsibilities, donned a more dapper look, sporting a mustache and often dressing in a suit, tie, and vest. Very early on, drawing on his experience as an undergraduate, he tinkered with the idea of a dissertation that centered on the ecology of parasites, but no one seemed interested. As he searched for alternative thesis topics, Calhoun took Orlando Park's class in field zoology. Never shy to share his thoughts on field work or zoology, Calhoun impressed Park and was in turn taken by his professor's "willingness to be influenced by a young, green graduate student."

John Calhoun, circa 1940, as a PhD student at Northwestern University. Photo credit: Catherine Calhoun.

Park had received his PhD across town at the University of Chicago and was well known for his work on the ecology and population dynamics of insects, particularly beetles. A field biologist to the core, Park spent many of the summers immediately before Calhoun began at Northwestern studying the nocturnal ecology of beetles living in the forests of Barro Colorado Island in Panama. Park had learned much from his literary hero, Sherlock Holmes. Like Holmes, Park was known for his attention to the smallest of details as he traversed the night forests of Barro Colorado. In time, Park paid his hero back, eventually editing *Sherlock Holmes, Esq., and John H. Watson, M.D.: An Encyclopedia of Their Affairs*.

Park and Calhoun soon grew close: close enough that Calhoun sat in on sessions of Park's Dixieland band, the Academic Cats, who played impromptu gigs in Park's apartment and at the Chicago Academy of Sciences: "It was common for a pedestrian walking by the Academy on a Wednesday," one of Park's colleagues wrote, "to hear strains of 'Muskrat Ramble' floating out of its windows." But mostly Park and Calhoun discussed ecology, field biology, and population dynamics, and soon Park took on the role of Calhoun's thesis advisor. And that thesis was beginning to gel around population dynamics.

As Calhoun reviewed the leading journals, he could see that population dynamics was a topic of growing interest to ecologists. During his years at Northwestern, even a cursory read of just the titles of new articles appearing in journals like *Ecology*, the *Journal of Ecology*, *Journal of Mammalogy*, *Physiological Zoology*, and the *American Naturalist* would turn up the word "population": titles included "Winter Reduction of Small Mammal Populations," "Home Ranges and Populations of the Short-Tailed Shrew," and "Population Studies in Colonies of *Polistes* Wasps," as well as a symposium on "Population Problems in Protozoa" published in the *American Naturalist*. Probing deeper into the articles themselves, and not limiting himself to papers with "population" in the title, Calhoun would come across hundreds of studies on various aspects of population growth and decline.[3]

For Calhoun's embryonic dissertation, the burning question became, The population dynamics of what? That remained an open matter until somehow, likely through connections that Orlando Park had made at the Smithsonian Institution field station on Barro Colorado Island, Calhoun met Alexander Wetmore, the secretary of the Smithsonian. Wetmore offered Calhoun a six-week internship in the summer of 1940 trapping small mammals around Reelfoot Lake, Tennessee, which largely meant trapping rodents of one sort or another.

Calhoun began the project by adopting a well-used approach to trapping at the time. The method involved laying out "long meandering lines of traps across each type of habitat and removing them after the catch began to dwindle" in order to sweep a large area before moving on. But the more he thought and read about that approach, the more frustrated he became: "I can remember when the thought struck me, 'What an inept way to learn about the range and population characteristics of small mammals.'" That approach, he reasoned, might work for collecting specimens for museums, which was, in fact, what the Smithsonian was paying him to do, but it ignored the ecology of the organism: what it ate, where it lived, how it behaved, and more. He searched for a more systematic approach that would serve not just collection managers but also shed light on the ecology of small mammals. Scanning the literature, Calhoun came upon a recently published 150-page paper on what's known as grid sampling, and that paper, together with some ideas on sampling that Raymond Pearl had published, converted Calhoun to this approach. Grid sampling, which Calhoun used for the latter part of his internship and is still used by many ecologists today, entails laying a systematic grid (not "meandering lines") of traps and sampling them at set, predetermined intervals. This provided Calhoun with his first fleeting, six-week glimpse of how rodent population sizes change over time.

Calhoun was more than comfortable working at Reelfoot Lake in the northwestern corner of his home state of Tennessee. That summer of 1940, he set out grids of traps in forty-two areas on Goat

Island, on the flood plains of a local bayou, and on nearby farmlands. Traps in each area were baited with balls made of oats, peanut butter, dried bread, raisins, and persimmon seeds, held together, in part, by some olive oil added to the mix. These traps captured almost two hundred small mammals from seventeen species, the vast majority of which were mice (white-footed mice, cotton mice, house mice, and meadow jumping mice), rats (marsh rice rats), voles (prairie voles), and shrews (short-tailed shrews).

Calhoun was more concerned with population sizes than with the number of rodents that happened to fall for his bait, and he used the trapping data to estimate the relative size of populations of rats, voles, or shrews in a particular habitat. What's more, Calhoun sacrificed the trapped animals to analyze their stomach contents. This allowed him to document a key component of the ecological niches in all those rodent species: dietary preference. He'd never done anything remotely like this work and found himself enamored with both the process and the product. "By the end of that period," he recalled, "I had decided that for my doctoral thesis, I would study the population dynamics of rodents." He settled on looking at the population dynamics and the twenty-four-hour rhythm cycles of the prairie voles he had worked on during his time at Reelfoot Lake and of the hispid cotton rat.

Straightaway Calhoun wrote A. J. Lotka, who had inspired Raymond Pearl to do his work on logistic growth, for his thoughts on such a project. A second-year graduate student, Calhoun was delighted that Lotka took the time to write back: "I was much interested in what you tell me of the work you are undertaking," Lotka wrote Calhoun. "I shall be most interested in how your enterprise progresses." In terms of day-to-day feedback, in addition to the advice of his thesis advisor, Orlando Park, Calhoun had a cadre of ecologists and animal behaviorists he could turn to twenty miles away, an hour-plus train ride on Chicago's purple and red "L" trains.

"At this time of intellectual searching, I had the great fortune of being exposed to the 'Chicago School of Ecology,'" Calhoun wrote, "during the time Allee, Emerson, Park, and Park and Schmidt were writing their monumental [textbook] *Principles of Animal Ecology*." Using the first letter of the authors' last names, the book soon became known as AEPPS (pronounced "apes"). The first Park on the list of AEPPS authors was Calhoun's advisor, while the second was Orlando's younger brother, Thomas, an ecologist at the University of Chicago whose work on the population dynamics of flour beetles soon became a classic in ecological circles. Right from its opening pages, population looms large in *Principles of Animal Ecology*. The authors wrote in the introduction, "We view the population . . . as a biological entity of fundamental importance. This entity can be studied with some measure of precision, and the emergent principles are significant throughout the field of ecology."

In the late 1930s and 1940s, ecologists were still a relatively small, close-knit lot, and there was a lot of, arguably far too much, academic nepotism. Graduates from a PhD program were often snapped up by the university that granted them their degree. Warder Clyde Allee, the de facto leader of the Chicago School of Ecology and first author on AEPPS, had served as PhD advisor to both Orlando and Thomas Park, making him Calhoun's grand-mentor. Few people would impact Calhoun's early career development more: "I fell under the influence of Professor W.C. Allee," Calhoun wrote of his days at Northwestern. Allee shaped Calhoun's philosophy of science, opening his eyes to "the possibility of fusing concern with social problems and ecological studies."[4]

Allee received his PhD from the University of Chicago in 1912, and in 1921, in another example of academic nepotism—the Zoology Department was quite pleased with itself—he returned to the university, now as assistant professor of ecology and animal behavior. Bucking the consensus that overcrowding always has deleterious effects on animals living in groups, Allee began a series of experiments examining whether there were also detrimental effects of *undercrowding*. He hypothesized that in undercrowded conditions individuals might miss out on the chance to reap the benefits of cooperation with other group members. To test that idea, Allee ran a series of experiments in his laboratory and at the Woods Hole Experimental Station near Cape Cod, Massachusetts, which documented a plethora of benefits that increase with group size—longer life, better protection from predators, reduced heat loss, and increased fecundity. Many of these studies were gathered in Allee's 1931 book, *Animal Aggregations: A Study of General Sociology*, which sat on "the border-line field where general sociology meets and overlaps general physiology and ecology."

After establishing that increasing group size could be beneficial to group members, Allee set to work to show that underpopulation might, under certain environmental conditions, rob individuals of such benefits. From his time at Woods Hole, he knew starfish were rarely found close together; instead, they lived a solitary existence, and they used eelgrass to hide from predators. With eelgrass around, small populations of starfish, interacting only on rare occasions, did just fine. But when Allee brought starfish into his laboratory and placed them into small dishes containing only seawater and no eelgrass, the starfish formed large clusters, packed so tightly that it was difficult to tell one individual from the next. When Allee added eelgrass to the dishes, the aggregations dissolved, and starfish went their lone ways. When eelgrass was available, starfish were solitary because the eelgrass protected them from predators; when the environment changed, and they lacked protection from the grass, they came together in what Allee interpreted to be a simple form of cooperation.

This protocooperation set the stage for more complex cooperation in more complex creatures. Allee came to think that this transition from protocooperation to cooperation was not only important in and of itself but also for what it tells us about our own behavior. "Even though one should admit that the proper center of the study of mankind is man and his work," Allee wrote in a magazine article, "many of us are convinced that without a knowledge of general sociology we are likely to regard the social traits exhibited by man as being peculiarly human when many of them are merely human variations of social traits common to animals in general." "Many of us" would soon include John Calhoun. But not yet.

In 1938, the year before Calhoun started at Northwestern, Allee published *The Social Life of Animals*—later rebranded under the title *Cooperation among Animals with Human Implications*. That same year Allee gave a series of six lectures based on that book at Northwestern, but Calhoun was first exposed to this idea of animal sociology and its implication for humans when, in graduate school, he began attending weekly Sunday afternoon gatherings of Chicago ecologists in Allee's home at the edge of the University of Chicago campus. Orlando and Thomas Park were permanent fixtures at these meetings, and there was a rotating bevy of leading ecologists from the University of Chicago, Northwestern University, and the Field Museum of Natural History presenting their ideas to one another, as well as to dozens of graduate students who sat in. Calhoun recalls lively discussions about population dynamics, including the perils of overpopulation and underpopulation, as well as how animal ecology could inform human societal issues. After sitting in on these meetings, Calhoun thought of himself not just as Orlando Park's student but more generally "under the tutelage of several mentors within the 'Chicago School of Ecology,'" and none more so than Allee.[5]

Under that tutelage, Calhoun began mapping out his dissertation work on the population dynamics and behavioral activity patterns of prairie voles and cotton rats. Among other things, he would study aggression, mating patterns, and maternal care, and look at how those behaviors changed as population size increased during a two-year experiment. To have some control over experimental design, he planned to do that work in large enclosures in which populations could live in seminatural conditions. Designing the experiment, he tapped into all he had learned from his many advisors, local and beyond, by, as Calhoun described it, "looking at social relations . . . (Allee's influence), examining population dynamics (trying to do what Raymond Pearl did with fruit flies), and recording the activity cycles in this semi-natural condition (O. Park's . . . influence)."

No sooner had Calhoun settled on the basic design and questions for his dissertation than he had to abandon a key component of the work. "World War II necessitated a drastic change in my plans," he wrote, with his graduate teaching responsibilities shifted to a sped-up premed program to produce much-needed doctors. As much as he wanted to, it simply would not be possible to sequester himself in the forest for two years, spending his days studying what was happening in his rodent enclosures. Instead, he would have to do a scaled-down laboratory version of the work. One unintended consequence of spending more time on campus was that it gave Calhoun's relationship with Edith Gressley, a sophomore biology major he was dating, a chance to blossom. Midway through Calhoun's time at Northwestern, the two were married.

In Calhoun's new thesis project, prairie voles and cotton rats, many of which Calhoun caught in the woods of Tennessee, were placed in small arenas measuring six feet long, three feet wide, and three feet high, and in which temperature, light cycles, and the availability and amount of food were all controlled. He focused on the behavioral activity patterns of the rodents: looking at peaks and troughs in activity over the course of twenty-four-hour periods, whether these peaks and troughs were the result of an internal clock that triggered various behaviors, and whether factors like temperature and the hours of light per twenty-four hours changed activity patterns.

To do all this Calhoun constructed an elaborate set of devices, including a "rodent activity monitor" hooked to a wheel that a rat or vole could enter and spin around. Every 140 revolutions of the wheel printed a line on a piece of paper, producing what looked like the printout of an electrocardiogram (EKG). What's more, he constructed a maze of sorts with electronic triggers that when tripped would ring a bell and punch a hole on recording paper. Calhoun also made direct observations on what was happening in his rodent microworlds, in which the rats or voles spent days, sometimes weeks. But the laboratory nature of the work severely constrained the population dynamics component of Calhoun's dissertation. These were not natural populations that would increase and decrease in size as in the planned enclosure work; rather, they were groups set at a fixed number and studied for relatively short amounts of time. As a partial remedy, and to get some sense of how population size affected activity patterns—especially aggressive behaviors—Calhoun experimentally manipulated group size. In one set of trials, a lone rat or vole was tested, and in others, groups of two, three, four, six, or seven rats or voles were tested together.

In August of 1943, John Calhoun stood before his dissertation committee, as well as students and faculty of the Department of Zoology at Northwestern, to defend his PhD thesis. With the assistance of a seemingly endless series of slides filled with EKG-like printouts, and graphs flashing up on the screen behind him, he stepped through his results. Both the rats and the voles tended to be much more active at night and increased temperature reduced their activity levels. Those results were standard fare in ecology and would not have surprised many in the audience that day. A bit more surprising, but not shocking, were his findings that reducing the amount of food led to animals being more active during daylight hours and that by shifting the light cycle in the lab by twelve hours, Calhoun could shift the internal clocks of the rats and voles by twelve hours.

Most interesting to both Calhoun and his audience were the complex ways that population size affected behavior. Increasing the population from one to two to four didn't change *when* rats or voles became active, but it did triple their activity levels. Competition in groups of four led to increased rates of aggression compared to competition in pairs, though not dramatic increases. But once the population was greater than four rats, violence escalated dramatically, particularly when the number of refuges (in the form of tin cans) was less than the number of rodents: "In the seven days from July 31 to August 7, 1942," Calhoun wrote in the *Ecology* paper that came from his dissertation work, "when six males were present but when only four nest cans were available competition became severe. Several rats received bad lacerations on their sides and one had his tail bitten off."

Calhoun expected competition to increase with group size, but the extreme aggression caught him off guard. At this early stage in his career, he couldn't explain it but made a commitment to himself "to try to understand the interrelationship between social behavior and population dynamics."[6]

3

THE PRIVATE LIVES OF RATS

The middle of World War II was an inauspicious time for a recently minted PhD in ecology to find work. Calhoun kept an eye out for open positions in academia, but he also put out feelers to anyone and everyone who might be able to help him land a job. When he was offered a one-year position at Emory University in Atlanta, the recently married Dr. Calhoun accepted: soon, Mrs. Calhoun found work as a medical technician on an avian malaria project at Emory. John Calhoun's job running the comparative anatomy labs had little to do with ecology but instead was part of a navy premedical training program. But the workload was relatively light and that gave Calhoun the chance to pursue ecology projects on the side. With no access to lab space or the funds do any enclosure work, and with on-campus teaching that meant long stints in field work weren't an option, Calhoun needed to get creative about what sort of ecology projects might be feasible.

He settled on a project that took him back to his teenage days in Tennessee. "Genetic adaptation intrigued me," he wrote. "After several false starts, an idea struck me from bird-banding days." He knew that English sparrows had been brought into North America a century earlier—with at least eight independent introductions prior to 1880. The sparrows quickly spread all over North America, into environments that ranged from the bitter winters of the north and west to the warmer environments of the south. Calhoun hypothesized that a century should be sufficient time for natural selection to operate differently on sparrows living in dramatically different North American climes.

Calhoun focused on wing length—a good predictor of overall body size—because estimates of wing length are easy enough to make from skins that museums keep in their collections. Through his connections in the birding world, and through a growing network of ecologists he was now part of, Calhoun was able to obtain sixteen hundred sparrow skins on loans from thirty-six museums: each skin had a tag with data that included where each bird was caught and when. Fortunately for Calhoun, but not so fortunately for the sparrows, he also wrote forty-two birder friends, who agreed to go out and capture a total of six hundred live birds. Using cardboard boxes Calhoun sent their way, these birders shipped the live sparrows to Calhoun, who then sacrificed all the birds. For some of those birds, he measured wing length from the skin; for others, he did direct measurements of wing bone length.

What Calhoun found was that the average wing length had increased significantly since the birds had been introduced a century earlier. More important, the data showed how natural selection led to this change. In a paper that emerged from this, Calhoun told readers that, for a plethora of reasons, ornithologists had long thought that larger size was strongly favored in colder environments, and his sparrows confirmed this relationship: from his data on twenty-two hundred sparrows, he found that as one moved across the country, from warmer to colder zones, the average size of sparrows did indeed increase.

The sparrows were Calhoun's first deep dive into evolution and natural selection. Up to this point, he'd been an ornithologist, mammologist, and ecologist, but he had never done evolutionary research per se. Now he had, and as part of the project, he immersed himself in the evolutionary literature, reading Julian Huxley's book *Evolution: The Modern Synthesis* and Ernst Mayr's book *Systematics and the*

Origin of Species, grounding himself in how natural selection operates in real populations.

Calhoun's position at Emory ended when the navy terminated the contract for its premedical training program in the summer of 1944. He sent out more than a hundred inquiries but got only one bite: from Ohio State University. Largely based on Orlando Park's glowing letter of recommendation, Larry Snyder, the chairman of the Zoology Department at Ohio State, hired Calhoun as an instructor for the large introductory biology course and lined up a medical technician job for Mrs. Calhoun in a hematology lab.

The instructor job, along with Edith's new position, put food on the table, but like at Emory, Calhoun's university duties were strictly teaching, with no funds or space for research. "For me, teaching the mass-standardized introductory course," Calhoun recalled, "did not prove to be particularly exciting." And so, again, he needed an outlet for his creativity. He sat in on a seminar group in genetics that Snyder ran. There, Calhoun started working on a project looking at the genetics of coat coloration in foxes. He tapped into already-published data about a pair of genes that code for red coat color versus black coat color in foxes that the Hudson's Bay Company had been keeping for two centuries, and he combined that data with the well-known nine-to-ten-year population cycles in foxes, hoping he'd find something interesting. He did—at least in a few of the populations he had data on. In those populations, the gene associated with black coloration decreased when populations were getting larger and increased when the populations were getting smaller. For Calhoun, the overarching importance of this study was not so much the details about the genetics of coat color but rather that it "left [him] focused on population cycles."

As at Emory, there was no time (or money) to set up and run a long-term field experiment, but there were populations of doves and robins that lived near the university arboretum behind the Biology building, and at least Calhoun could do something with living, breathing animals. The question he decided to address was run of the mill—what type of artificial nesting substrate did the birds prefer? It turned out that doves preferred black-colored nesting material and robins preferred green-colored material. "Scientifically this wasn't a very important effort," Calhoun recalled, "but it did show me that a minor change in the environment can drastically change the behavior of members of a population." Equally important, the birds led him back to rodents. Rats were feeding on the eggs of both the robins and doves he was studying, and to understand what was happening in the bird nests, Calhoun needed to understand the rats that were causing havoc there. He found a colony living in a shed near the arboretum, and to see how they were getting the eggs, Calhoun did some tinkering with the rat population: "Watching the rats, and trapping and releasing them," he wrote, "began to open for me a vista to the lives of these quite social animals."

When the year-long lectureship position at Ohio State was at an end, Calhoun found himself looking for work again. The ecology job market at universities was still very tight in 1945, so Calhoun cast his net broadly. He heard from a friend at the University of Washington that the Rodent Ecology Project at the Johns Hopkins School of Hygiene and Public Health was looking to add a new member to study Norway rats in the row-house neighborhoods of Baltimore. Rats were a perennial problem, in terms of both basic sanitation and disease transmission, and they had a predilection for the often run-down, economically depressed row-house neighborhoods of downtown Baltimore. The Johns Hopkins team, funded by the Rockefeller Foundation, the City of Baltimore, and the Office of Scientific Research and Development in Washington, DC, was one of many groups working on how to rid these neighborhoods of rats.[1]

Between 1930 and 1940, the Baltimore metropolitan district grew more than 10 percent, to slightly over a million people. The housing that already existed in the district, as well as the new housing built to accommodate growth, was often subpar: the 1940 census found that thirty-four thousand dwellings in Baltimore lacked flushing toilets.

A rat hugs Lord Baltimore as garbage accumulates on the street. Baltimore's health commissioner stands in horror in the background. Note the giant rat on the Washington Monument. Cartoon by Richard Yardley. "Man of the Year," *Baltimore Sun*, December 31, 1942. Reproduced by permission from Baltimore Sun Media. All Rights Reserved.

Many of those dwellings were row houses (houses connected by common sidewalls). Most were two or three stories tall, and alleys often separated parallel strips of row houses. Those alleys, and the garbage deposited there, were prime real estate for rats.

The blight of the most rat-infested neighborhoods, largely inhabited by low-income, often African American families, went on display for all when the *Baltimore Sun* ran a series of articles, based largely on social activist Frances Morton's detailed study of housing in the city's Wards 5 and 10. In response, the Baltimore City Council, under intense public scrutiny, passed the Hygiene of Housing Ordinance in 1941, which in time morphed into what became known as the Baltimore Plan—the first large-scale attempt in the United States to rehabilitate entire neighborhoods. That plan included, among other things, provisions and funds for the serious rat problem plaguing the neighborhoods.

The Rodent Ecology Project, a descendant of the Rodent Control Project, was the brainchild of Curt Richter, who started the latter project in 1942. By the time Calhoun applied for the new job, it was David Davis who was running the operation, but Richter was still a key player, so Calhoun read up on Richter's and Davis's interests and found they overlapped with his own. Each had done a great deal of work on rodent behavior, with Davis being the more ecological of the pair, but Richter shared Calhoun's early interest in parasites and disease. "Soon I found myself giving a seminar before a small selection committee [at Johns Hopkins]," Calhoun recalled, and that committee, impressed with both Calhoun and his body of work, offered him a job. So, the Calhouns moved again, this time to a house in Towson, a suburb a thirty-minute drive north from the Baltimore row houses and their rats.[2]

Little early on suggested Curt Richter would be a biologist, let alone lead the Hopkins Rodent Ecology Project. As a Harvard undergraduate, he started off as an economics major. One early course he took focused on John Stuart Mill, David Ricardo, and Adam Smith. Six weeks later his professor sent him a note: "Mr. Richter, I do not think that economics is your forte. Please drop the course at once." He did. Floundering, he took advice from some friends that led him to a class on animal behavior taught by Robert Yerkes. "This course, though short," Richter recalled, "made me feel that at last I had found something that really interested me." Richter soon was reading everything he could on behavior, and in 1919 he entered the PhD program at Johns Hopkins, working on the behavior of rats in the lab of John Watson, a leader in the early study of animal behavior. After he finished his dissertation work, Richter was immediately granted a professorship at Johns Hopkins and continued to work on rat behavior, including feeding behavior.

In the late 1920s and through the 1930s, Richter did a series of studies that found that laboratory (white) rats were remarkably good at distinguishing food sources based on their nutritive value. Soon he was running experiments on taste and odor discrimination in rats, including whether they could discriminate foods that had been baited with poisons. By the early 1940s, with the rat problem looming larger on the streets of Baltimore, understanding what rats could taste and not taste began to take on consequences far greater than the academic interests of a professor at Johns Hopkins. Rats tend to avoid foods with novel smells, and that meant a good rodenticide was one that rats couldn't smell when it was dusted on food as bait.

It wasn't just the citizens of Baltimore who cared about such matters: a 1942 report estimated the annual cost of rodent damage in the United States at $200 million. The Defense Department in Washington also had a stake in understanding rat taste discrimination and rodenticides. With World War II burning hot, there were military leaders who feared the Germans might use rats as a vector to introduce disease into the cities of America, and so the development of tasteless, odorless rodenticide had national security implications.

Richter started off working on rodenticides using a chemical called carbamide, in part because a great deal of work had been done on human taste and carbamide. When a very small (nontoxic) amount of carbamide is soaked on a piece of paper and the paper is dried and placed between the lips, some people sense a bitter taste, but others taste nothing at all. To see how carbamide would affect wild Norway (brown) rats on the streets of Baltimore, Richter needed help, as he and his team of laboratory scientists had no experience at all with Norway rats. A sanitation worker told Richter that he would find all the rats he needed and more around grocery stores and near the city dump. "I shall never forget this first introduction to the Norway rat," Richter wrote in an autobiographical piece he titled *Experiences of a Reluctant Rat-Catcher*. "This officer was a young man dressed in a very natty blue uniform . . . we saw a number of rats scurrying [at the dump] . . . he dashed down a slope, leaping about like a young gazelle over old cans and piles of rubbish in pursuit of a rat. He chased it from under one can or box after another; then suddenly I heard a loud squeal as he held the rat up in his hand. I never found anyone to duplicate this performance." In no time, Richter had all the street rats he needed.

Richter found that, when ingested, carbamide was toxic to the Norway rats roaming Baltimore. In terms of using it as a potential rodenticide, the problem was that the rats could smell carbamide, and they quickly learned to avoid it. Richter then began tinkering with chemicals in the same family as carbamide and eventually produced a compound, ANTU (alpha-naphthylthiourea), that was one hundred times more toxic than any other rodenticide available at the time and that had only a weak odor. The question was then the best way to bait traps with ANTU. Richter selected eight city blocks near the Johns Hopkins Hospital and somehow corralled eight Boy Scouts into doing the baiting. The Scouts placed piles of ground-up corn mixed with ANTU or corn cobs dusted with ANTU in yards and alleys, as well as in cellars of the row houses. Both baits worked well, attracting and killing rats.[3]

Richter's work on ANTU led to the birth of the Rodent Control Project. In October of 1942, Baltimore's mayor, Howard B. Jackson, provided $5,000 to hire a team of workers, and the Office of Scientific Research and Development in Washington, DC, provided ANTU and bait mix for Richter and his team to run a large-scale trial spanning two hundred city blocks near the Johns Hopkins Hospital. As with the smaller-scale, eight-block study, results were encouraging, with kill rates often in the 80–95 percent range. Mayor Jackson was impressed enough to increase his office's funding of the work: "There are no strings attached to the appropriation," Jackson told Richter. "Anything you suggest will be done. We are giving you $25,000, and all we ask is that you kill $25,000 worth of rats." Over the next few years, the City of Baltimore would use one hundred thousand pounds of ANTU-poisoned bait to deal with rats infesting 4,160 city blocks. But thousands of blocks, and indeed even hundreds of blocks, comprised too large an area for the Rodent Control Project to do long-term detailed observations on rat populations, and so Richter selected a twenty-eight-block area near Johns Hopkins, one of the most blighted areas of the city, for that more detailed study.

Instead of Boy Scouts, Richter and his team at the Rodent Control Project worked with "fifteen young, enthusiastic air-raid wardens in the area [who] volunteered for this job." Each week the team would focus on a subset of the twenty-eight-block area. On Monday or Tuesday, they would post signs and distribute flyers that the block was going to be baited with ANTU that coming weekend. The flyers asked all residents of the block to clean their yards and cellars and put out garbage for pickup by the Sanitation Department on Saturday morning. Then on Saturday the air-raid wardens would place out yellow corn meal mixed with ANTU or corn on the cob mixed with apples

A typical Baltimore city block like those studied by the Rodent Ecology Project. Credit: Broadway Slum Redevelopment; Citizens Planning and Housing Association Records; series 8, box 1, folder 21; University of Baltimore Special Collections and Archives.

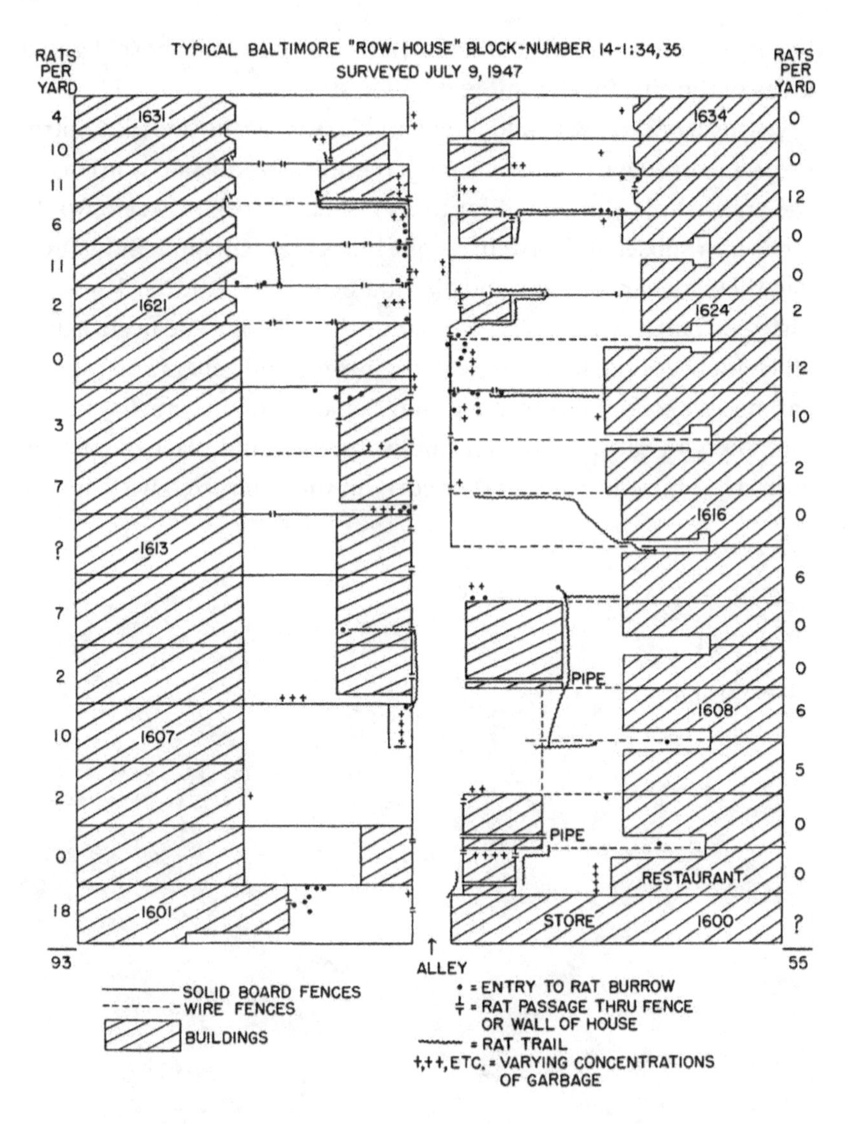

One of many schematics made by the Rodent Ecology Project team. This schematic is from study block 14 from a survey on July 9, 1947. Credit: US Department of Health and Human Services.

and sweet potatoes, all lightly dusted with ANTU. On Sunday morning the air-raid wardens returned and collected the dead rats. "Always an exciting occasion for all workers," was how Richter described those eerie Sunday mornings, with sometimes hundreds of rat bodies collected and likely more hidden and forgotten.

By the time Richter stepped down as leader of the renamed Rodent Ecology Project in the winter of 1944, across the larger citywide program of more than four thousand city blocks, close to a million rats had met their demise by ingesting food from ANTU-baited traps. In that sense, the project had been extraordinarily successful. But there was a problem. Within months after a population on a block was culled by ANTU traps, with kill rates between 50 to 90 percent, the population would return to its earlier size: ANTU was a phenomenally good predator of sorts, but apparently there were other ecological forces in play.

To take a more explicit ecological slant, shortly before Richter stepped down as leader of the Rodent Ecology Project, he recruited John Emlen, an expert on animal population dynamics, to take the helm. Emlen began censusing rat populations using modified rabbit traps. He would mark the rats, release them, and try to recapture them to gauge the size of a rat's home range. He also fed some rats a blue dye and traced their movements by following their droppings. Unfortunately for the Rodent Ecology Project, others also recognized Emlen's talents as a population biologist; not long after he came on board, he accepted a faculty position at the University of Wisconsin, and that's when ecologist David Davis took over the Rodent Ecology Project.[4]

Davis and Emlen's ecological approach added several dimensions to tackling the Baltimore rat problem. For one thing, both wanted to understand what, from a rat's-eye perspective, constituted a population: a group of individuals that interacted with each other but not with others. What they found was that rat populations mapped remarkably well on Baltimore city blocks, with each block home to a population of about 150 rats. Each block was, Davis and his student

John Christian noted, "effectively an island and its rats form a discrete population unit since immigration and emigration of rats is negligible or absent."

It was Davis who taught Calhoun the ropes for working in the rat-infested neighborhoods of Baltimore. Right from the start, Calhoun was fascinated with the fact that after a block that was home to about 150 rats was decimated by ANTU, the population rebounded: "What was a block-box nearly devoid of rats," Calhoun wrote, soon "reached the prior [number] of about 150 rats." But not more than 150, and that was the intriguing part. From his work on rodents and birds in the field, Calhoun knew that in the wild, food was always a limiting factor, but on a Baltimore city block, what he saw, day after day, was that "yards usually contained abundant sources of food in open garbage cans." There was enough food, he thought, to sustain far more than 150 rats, and so the availability of food did not act as a break on population growth: in the language of ecology, food was not limiting the carrying capacity. Perhaps, Calhoun began to think, rat populations were self-regulating, well below the carrying capacity of their environment. But how?

To answer that question, Calhoun found himself "climbing high board fences in the backyards of Baltimore delving into the lives and times of wild Norway rats." He captured scores of rats and marked each by attaching numbered tags to their ears, released them, recaught them when lucky, released them again, and repeated the process over and over to better understand the population on that block. Because of his earlier work on voles, hispid rats, chimney swifts, robins, and doves, Calhoun did his best to also watch the rats interacting with each other to catch a glimpse of what he described as "the private lives of rats."[5]

The more time Calhoun spent on the streets of Baltimore, the more he realized that social behavior, which to that point had largely been overlooked by the Rodent Ecology Project, might be important in understanding population growth in rats. His mark and recapture work confirmed Emlen's and Davis's findings that city blocks really did demarcate rat populations; rats virtually never migrated from one block to another. For all practical purposes, each rat had all its behavioral interactions with a relatively small number of others on its own block—and with no other rats. Those clannish behavioral dynamics might very well have had something to do with why populations tended to settle at population sizes well below what the food on a block could sustain.

The only way to know was to set up an experiment.

4

RAT MAN

As Calhoun's thoughts on an experiment to understand the population dynamics of rats began to gel, he made a strategic decision. Aside from one small pilot study done by Davis, all the work by the Rodent Ecology Project involved using ANTU to *decrease* population size and then measure what happened. Calhoun decided to come at the problem from the other end and "override . . . rat clannishness" by establishing a "second layer of rat society in an already established society." The idea was to embed that new layer within "the less used interstices of [that] established society." All of which is to say that to understand why rat populations seemed to settle at about 150 (far below what a city block could sustain), rather than decimate the population with ANTU, Calhoun would increase population size and see if it headed back down, and if it did, he would try to determine why.

Calhoun's experiment involved three adjoining city blocks that he labeled the northern, central, and southern blocks, each of which had a population in the 100–150 range. "Rat man," as Baltimore residents on those blocks came to call Calhoun, first trapped and then marked rats from all three blocks. He then introduced 112 marked alien rats onto central block, where he had marked 75 rats, and he used the northern and southern blocks as controls: nearby blocks where aliens were not added.

Calhoun used live traps in the central block to look at how the population of resident and alien marked rats was faring. As he did so, the resident humans on the block, who knew rat man but not the details of the experiment he was doing, told Calhoun that the rat problem on their block was getting worse, as he knew it would be given the design of his experiment. People naturally assumed Calhoun's goal was to get rid of rats, and they tried to be as helpful as they could. Sometimes that kindness backfired on Calhoun (and the rats). "They tried to help me," Calhoun wrote. "Thinking that I shared their belief that every good rat is a dead rat, every so often someone would gather up the traps containing tagged rats that I wanted to record and release, pour boiling water over them, and meet me smiling."

Calhoun's results from the central block were not pretty. Many rats were found dead, and data from who was and who was not found in live traps indicated the rat body count was even higher than the number of corpses Calhoun saw. Mortality was anything but random. Calhoun estimated that aliens suffered mortality rates three times higher than central block residents. As to whether the population started to drop back to preinvasion levels, what Calhoun discovered was that just a few days after the release of the aliens onto the central block, the total population there dropped 15 percent, on its way back to the preinvasion level of near 150.

The ANTU work had shown that if the 150 or so rats that were living on a Baltimore block were poisoned down to a much lower number, they soon began heading back toward 150, though there was food for plenty more. Now, Calhoun had shown that if the number of rats on a block was experimentally increased, it soon started heading back down to 150. That, in and of itself, was an important finding with respect to the dynamics of population stability, but Calhoun wanted to know *why* his experimental population started to decline so quickly. For that he had to make some educated guesses, which he was not shy about doing. Calhoun had only anecdotal information on the behavioral interactions between aliens and residents, but from what he did know from his observations, he thought that competition, in the form of dangerous fights between residents and aliens, was why "all hell broke loose" on central block.[1]

Calhoun was becoming convinced that "survival required some essential modicum of social stability" and that his introduction of alien rats had shattered that social stability, leading to violence and population decline. Davis was starting to see similar things when he looked at rats in the laboratory. "Psychological turmoil" was how Davis described what happened when aliens were added to a laboratory rat population with an already-defined social structure. When asked what that turmoil entailed, and how it would affect population growth, Davis told an audience full of biologists to imagine adding a very large "rat into a colony which, up to that time, was quite happy." A colony where "everybody knew who was married to whom, and whose children were whose." It seemed likely to Davis that "there would be a considerable amount of commotion" and that the new rat "would try to establish himself as the dominant." Davis thought it was "entirely possible that factors of that type would explain [subsequent population] decline."

Although Calhoun published his results on central block in a technical, rather specialized, journal called the *Journal of Wildlife Management*, this experiment marked the first time his work was picked up by the national press. "Rats are as bad as human beings in some ways," was how *Time* magazine opened its June 14, 1948, article, "Displaced Rats." What Calhoun was up to, the readers of *Time* learned, was studying "the troubles which refugee rats have to put up with when they emigrate." There was a plethora of those troubles, *Time* wrote, and "at once there was social strife." *Time* interviewed Calhoun for the article and summarized what rat man said: "In a long-established rat community, says Dr. Calhoun, there is very little fighting. Every rat, having tested its strength against its neighbors, knows its social position and stays in it. Newcomers must battle for places in this set order. But since (like human immigrants) they do not know the new country, they are at a disadvantage. When danger threatens, they do not know where to hide. . . . The native sons, familiar with local conditions, win most of the battles for position in the social order." Whether Calhoun actually said "country" and "like human immigrants" or whether *Time* was paraphrasing is not clear, but given the language in Calhoun's later writings, it is not a stretch to imagine the words were his.[2]

Results from central block piqued Calhoun's interest in the private, but apparently quite social (albeit dangerously social), lives of rats, and how that sociality affected population dynamics. The streets of downtown Baltimore provided Calhoun with a natural laboratory, but there was only so much control he could have there, because people lived in this laboratory. Residents killed rats in his live traps or moved the traps, and empathetic youngsters occasionally released a rat from a trap. What's more, block residents might take garbage out one day but not the next. They might put out rat poison when they deemed it necessary. They might tidy up the alley at an inopportune time, right in the middle of an experiment. Of course, Calhoun didn't begrudge residents the right to do any of those things and more, but they could make interpreting the results of experimental manipulations difficult.

To have the control he would need to truly understand the private lives of rats, Calhoun decided he would need to construct "a habitat that simulated the Baltimore row-house rat habitat" and to design a study "to determine the manner and extent to which social interactions might influence population growth." Davis and others on the Rodent Ecology Team agreed and gave him freedom to do just that. Suddenly, the huge amount of planning that he had put in for the almost-but-not-quite enclosure experiment he had wanted to do as part of his PhD back at Northwestern would yield fruit.

The first question for Calhoun was where to build this large rat world. His house in Towson, Maryland, was surrounded by hundreds of acres of fields and forest. In the fall of 1946, as John O'Donovan, owner of the surrounding property, was taking his dogs out for a walk, Calhoun joined him and asked whether he might build his experimental enclosure on a field a hundred yards from the Calhoun home. "Not knowing really what I meant by a 'rat pen,'" Calhoun recalled, "[O'Donovan] told me to go ahead and make use of his land. . . . After settling a few legal problems, he left me in peace to become a peeping Tom of the sex and other life of rats."

Calhoun began by sitting down and writing a list. He'd need to build an observation tower atop of which he could watch his rats. The enclosure had to be large enough that it could eventually house many groups of rats but small enough that, once he marked all the inhabitants, he could observe anything they did above ground. The enclosure would need to have a tall fence around it, with posts every ten feet and barbed, electrified wire at the top, so vertebrate predators could not enter. He would need to make the fence ratproof, so that the residents could not leave and strangers could not enter. He would need one location for food and water. Rats would need to be able to build burrows where they liked. Finally, he would need to divide the enclosure into four parts by yet another short fence, which would "enable the formation and delimitation of local colonies" but also have passages that allowed movement into the sections.

With that daunting to-do list in hand, Calhoun hired two laborers and started building. He settled on a hundred-by-hundred-foot structure—roughly a quarter of an acre—as that was about the size of some of the smaller Baltimore city blocks where he had done his work. As for the rodent dining room in the enclosure, Calhoun decided it should be twenty feet in diameter and surrounded by a shorter fence that allowed access from passages on the north, south, east, and west. The feeder would be stocked with an abundance of Purina rat chow (used to feed laboratory rats everywhere), and on occasion some garbage, so the rats didn't grow bored with the same menu every day; it and the watering hole were constructed so that a dozen or more rats could feed at a given time. To make the rats feel even more at home, Calhoun added nine nesting boxes in each corner of the enclosure.

The enclosure that Calhoun erected in a field adjacent to his house in Towson, Maryland. Photo credit: Catherine Calhoun and National Library of Medicine.

By January 1947, the basic structure was in place. Next came the bells and whistles that would allow Calhoun to monitor the rats even when he could not be perched, six-power binoculars in hand, atop the twenty-foot-tall tower built just outside of the enclosure. To do that, Calhoun got a company in New York City to donate recording devices that he could set so that whenever a rat broke a beam of light emitted by the recorder, it triggered a pen to create a mark on a roll of recording paper.

Calhoun didn't have the resources to film the rats, but the United States Army did, and they wanted some footage of his work to use in a film they were making about pest control and epidemiology. And so it was that Major Lown and Lieutenant Strickland, along with a team consisting of a dozen camera operators, a sound technician,

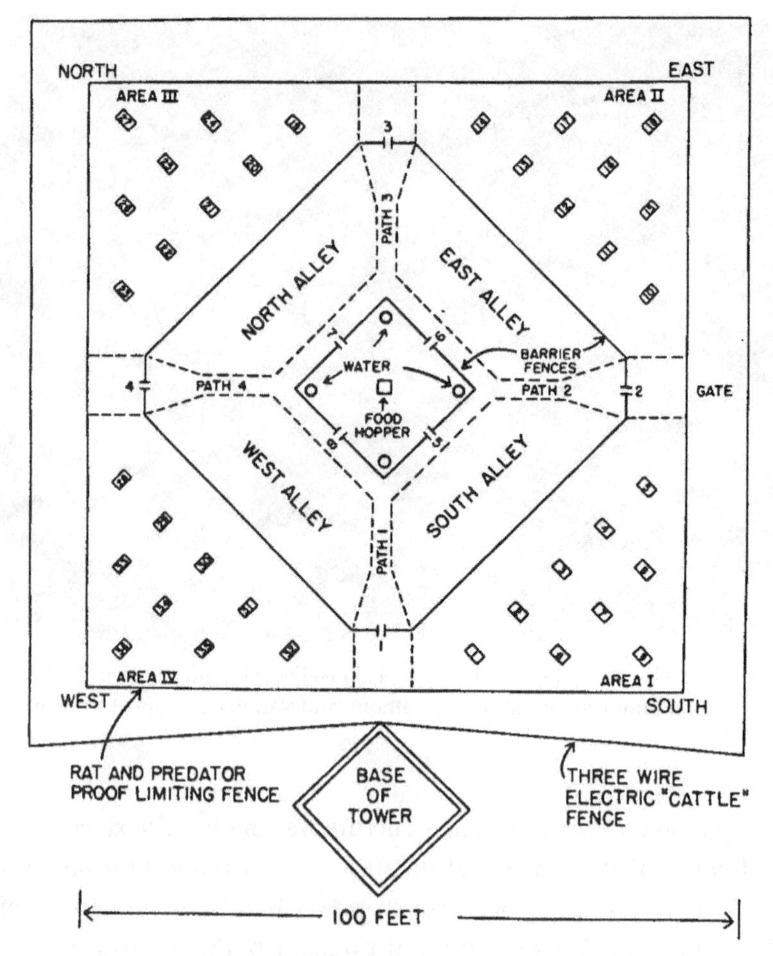

A top-down schematic Calhoun drew of the enclosure adjacent to his house in Towson, Maryland. The numbered rectangles are nest boxes. Credit: US Department of Health and Human Services.

electricians, and more, paid a visit to the Towson enclosure to do some filming. When they were finished, they gave Calhoun a copy of everything they had filmed, providing him details on the private lives of rats that he otherwise would not have had. Though the army movie was about rats as pests and disease vectors, after spending some time with Calhoun's rats, the director decided to have the narrator open with "the rat is a social animal" and then to close with a denouement of "the life of this so-called lower mammal is quite complex."[3]

Rats are largely nocturnal creatures, which posed challenges for watching their social dynamics. They started to get active around six o'clock in the evening, regardless of the time of year, which allowed Calhoun many hours to watch before summer sunsets but much less time in winter. To somewhat remedy that problem, he mounted powerful lamps over four sections of the enclosure (which the rats quickly got used to), using them to extend the time he could watch the colony from up on the tower.

As far as who would live in this large, relatively safe rat world, stocked with food and housing aplenty, Calhoun decided that the initial population would be small, so he could track growth and decline as it happened over the course of an experiment that would run for twenty-eight months. To obtain the original residents of the enclosure, on February 4, 1947, Calhoun took a ferry to Parsons Island, a 150-acre speck of land in the Chesapeake Bay where John Emlen and David Davis had done some work. Calhoun captured seven "wild" adult males and seven adult females. Those rats, and later their descendants—four rat generations when all was said and done—were marked with numbered metallic ear tags and by applying Nair hair remover to remove patches of fur from predetermined locations on a rat's back.

To gauge how population size was changing over time, every six weeks Calhoun captured every rat in the enclosure. To do this, he baited scores of traps with sunflower seeds, oats, cracked corn, alfalfa soaked in molasses, and slices of oranges and sweet potatoes: to apprehend any rats not drawn to traps, he excavated burrows and enticed rats out of the nest boxes he had added. During some censuses, Calhoun not only noted the identity of a rat but also anesthetized all the rats and gathered data on size, weight, sexual condition, and the size and number of wounds.

From the outset, Calhoun's overarching questions centered on population growth: given an environment that better mimicked the rat's natural one, but still allowed him some experimental control, would rat populations still settle down to numbers far below what the environment could, in principle, support? And if so, what role did social behavior play in the process? He made a rough, back-of-the-envelope calculation of how many rats could, in theory, live comfortably on his quarter of an acre (roughly ten thousand square feet) stocked with an essentially limitless supply of food and water. He needed some mode of comparison, so he asked himself how many rats could live in a lab space of ten thousand square feet, assuming each rat was placed in a standard two-square-foot cage. Calhoun thought that his five-thousand-rat answer was "actually a conservative one in regard to representing the biotic potential expected from this free-ranging [enclosure] colony." In fact, he thought "50,000 descendants . . . might have been alive. . . . Nevertheless, it is believed that the figure 5,000 is a more realistic one," and so that is what he used.[4]

As Calhoun saw it, the colonizers in the enclosure had to do two things, and quickly: "develop a social organization" and "learn a rather complexly structured environment." The rats learned where the food, water, and nest boxes were relatively easily, though four died in accidents, decreasing the initial population to ten. The key to the development of a social organization was the establishment of dominance hierarchies and group territories. The rats had been captured at different locations on Parsons Island, and in the absence of prior experience with one another, it took them time to sort out dominance hierarchies within and between groups, but eventually they did. In short time, the ten colonizers splintered into two groups: three males and two females who established their territory in the northwest section of the enclosure and two males and three females who set up residence in the southeast corner.

For the first year or so, the population grew slowly, from ten to thirty. Then a pronounced growth phase began, and over the next nine months, population size grew to about 170 rats. But then, as quickly as the growth phase had kicked into gear, it came to an abrupt halt. For the last six months of the experiment (December 1948–June 1949), population size fluctuated very little, never exceeding 180 individuals. "All the evidence is that this trend . . . would have continued at least into the next year," Calhoun wrote. "The best estimate that might be made is that by 1950 there would have been 77 breeding females."

Like the rats on the streets of Baltimore, despite having the food and space they needed, the rats in the enclosure settled at a population size well below what the environment could hold: indeed, the average number of young per female plummeted from nine in the colonizing generation to less than two by the closing days of the experiment. Harkening back to his back-of-the-envelope calculation of 5,000, and comparing final population size of around 180—remarkably close to the number found in a Baltimore city block of similar size to the enclosure—Calhoun asked himself, "[What] was the cause of this 25-fold decrease in utilization of space under naturalistic conditions?" Slowly, he began piecing together an answer to that question.

As with the initial colonizers, who splintered into groups almost immediately, rats throughout the course of the experiment were always part of demarcated group territories and generally stayed in their natal group for their entire lives. When Calhoun mapped group territories just a few months prior to the end of the experiment, four territories lay squarely between the fence around the food pen and the middle fence, in what Calhoun called the alleys; the other seven territories were, fully or in part, between the middle fence and the outer fence, further from the food pen, in corners of the enclosure where Calhoun had added nest boxes.

Rats in alley territories nearer to the rat dining room grew more quickly, and because size determines dominance rank, they tended to be dominant to those in territories in the corners. To access the food bins, corner rats usually had to pass through one of the territories in the alleys, and that meant trouble. Once rats were together at the feeder, there was remarkably little fighting: it was while getting to

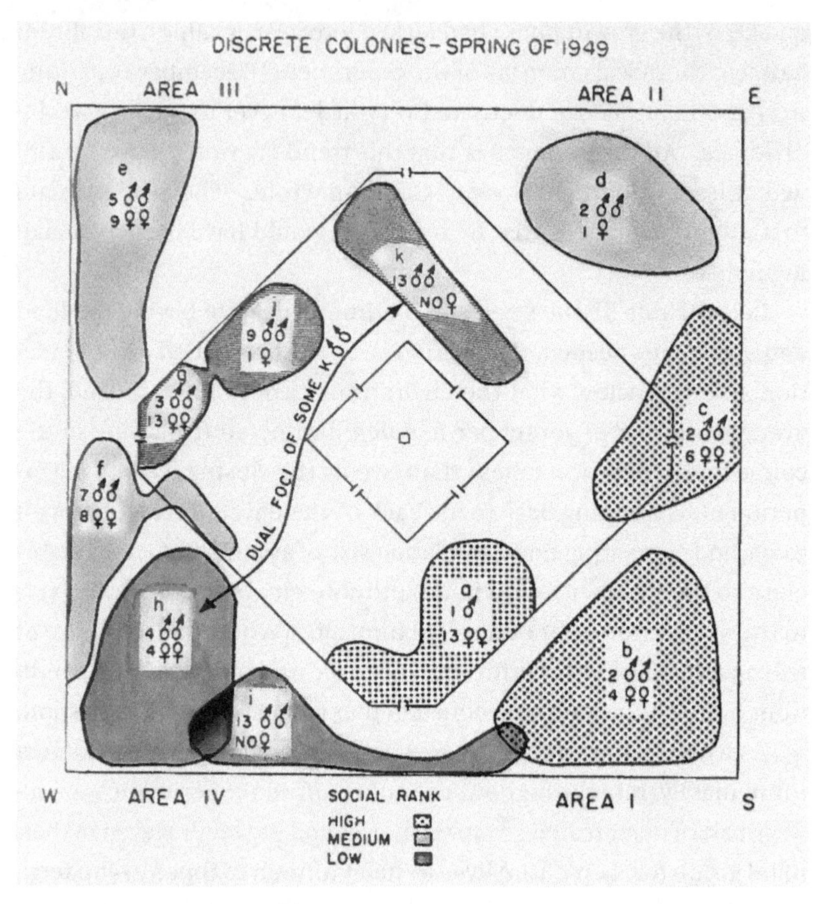

Eleven group territories were in place in the enclosure in the spring of 1949. The food hopper, as well as the water dispensers, were located at the center of the enclosure, within the diamond-shaped fence. Credit: US Department of Health and Human Services.

the feeder that the fighting took place, with rats in the alleys violently repulsing those from the corners who were trying to make their way in. Calhoun interpreted this difference as evidence that it wasn't food per se that caused fighting but rather that fighting increased when rats felt crowded. As he saw it, it was as if there was enough food for every rat to eat at the feeder without getting in each other's way and so there were no set rules to be violated. But when rats settled a territory, they somehow determined how many would be allowed in their group, and any number above that created a feeling of crowding that triggered aggression.

Rat fights can be nasty, with the losers having open wounds leading to lifelong scarring. Losers of fights grew more slowly than winners, and so, once rats from the corner territories started losing fights to alley rats as they tried to traverse their way to the food, the odds of them winning future fights got lower and lower. Losing repeatedly had devasting consequences: fewer offspring weaned (for females) and decreased chances of survival, largely because rats with open wounds from fighting were more likely to die from infestation by screw-worm fly maggots. And the cycle continued across generations. Smaller, less dominant female rats in the corners had less food to turn into milk for their offspring, and on top of that, they often displayed little maternal care, and so their offspring were small and weak. These offspring not only had to deal with alley rats but had to do so at a marked size disadvantage.

It wasn't just the interaction between alley and corner rats that was suppressing population growth: social stability, or rather the absence of it, did as well. "A stable group is one in which there is a well-developed dominance hierarchy," Calhoun wrote, "where there are well-established relationships between all the members of the group." The most stable and most productive groups were ones in which a single dominant male, or a pair of high-ranking males, lived with high-ranking females. Females in unstable groups were hounded by what Calhoun described as a "pack of males" following them around. On any given night, such females might be mounted hundreds of times by males roving about in packs. These females often built poor nests, had very low conception rates, and had young with high mortality rates, in part because they had been such poor mothers. Again, a vicious cycle was put in place, for not only did females in unstable groups have low conception rates and higher mortality among their offspring but those offspring that did survive produced fewer progeny of their own. As population size in the colony grew, the proportion of unstable groups grew along with it, putting the brakes on growth.

There was yet another force keeping population size below what it might otherwise have been. Though relatively rare, sometimes all-male groups formed. These were usually found in corners, rather than alleys, and were often made up of low-ranking males that had left another group because of the excessive violence they had experienced. "Most sex for them was homo[sexual]," Calhoun wrote. "They fed during the bright daylight hours when most of their nocturnal colleagues slept. They no longer retained the capacity to build burrows in the earth. Instead, they lived above ground . . . exposed to . . . [the] weather in winter." Even though Baltimore street rats rarely leave their home block, Calhoun was certain that on the streets such "outcast" males would have left to search a block where they had at least some chance, however small, of mating. But outcast rat males in the enclosure couldn't leave, and so Calhoun decided to call them "imcasts:" outcasts who had nowhere to go. As far as population growth was concerned, imcast males were just dead weight.

As Calhoun's work in Baltimore had hinted to him earlier, the complex, sometimes violent, social dynamics of rat life appeared to be what was limiting population growth, keeping it far below what the resources of the environment could sustain. Calhoun published some of his results in *Science*, arguably the most prestigious journal in the world for such matters, and spoke about the enclosure work not only at his academic home base of Johns Hopkins, where he "floated many trial balloons," but at a meeting of the New York Academy of Sciences, at the University of Michigan, and back at the University of Chicago. The tale of the Towson experiment was spreading in other ways. Calhoun had collected parasites from some of the rats in the enclosure and sent them to the National Institutes of Health (NIH), just down the road in Bethesda, Maryland, where he was slowly building connections. One day, three NIH scientists, including Carl Clausen, who was in the midst of creating the Laboratory of Socioenvironmental Studies at the National Institute of Mental Health, the newest of NIH's cluster of institutes, came by the Towson enclosure to see it for themselves. In passing, Clausen told Calhoun that he "could never fulfill the potentials this study promised unless such studies were conducted in some large indoor setting, more amenable to rigorous control of the physical conditions." One of the other visiting NIH scientists then added that plans at NIH included building such a facility at some point in the not-too-distant future.[5]

The rats in his enclosure, and what they taught him, got Calhoun thinking more generally about population growth, including in our own species. Even after the 1948 publication of two books on the perils of human overpopulation and limited resources—William Vogt's *Road to Survival* and Fairfield Osborn's *Our Plundered Planet*—in the late 1940s, the notion that population growth was a ticking time bomb had not yet fully gripped the public, or even much of the scientific community. It's not clear if Calhoun read Osborn or Vogt, but what is clear is that he *was* worried about the consequences of human population growth. "Human populations are approaching their maximum growth," he wrote in one of the papers that came out of the enclosure experiment. "If studies in animal ecology and behavior are to have greatest relevance to the understanding of human social and psychological problems, we must study the factors affecting populations at different stages of their growth and apply techniques of physiological and psychological analysis, developed for the study of laboratory animals under essentially asocial conditions, to the study of individuals taken from a social group whose past history is known." Species like rats.

It wasn't just that Calhoun was starting to think, write, and speak about the possible interdisciplinary implications of what he was doing—people from other fields were beginning to reach out to him about his work. As Calhoun was finishing up the Towson enclosure study, Jacob Marshak, an economist at the University of Chicago interested in boom-and-bust cycles, read a short write-up that the *New York Times* had published on one of Calhoun's New York Academy of Sciences lectures about the enclosure experiment. "There are certain methodological problems arising, which may be common to your work and our work," Marshak wrote Calhoun, "and I should be interested to learn a little more about your problems. Could you let me have some of your reprints on the subject?" Calhoun was more than happy to oblige; he not only sent copies of his articles but also told Marshak he would be giving a lecture on that very work at the University of Chicago the following month, adding that "the techniques of experimentation are amenable to the analysis of certain economic principles, as well as certain problems relating to cycles," though he didn't expound on which principles and problems. The two met the morning of Calhoun's lecture, and Calhoun clearly impressed Marshak as an emerging interdisciplinary thinker: in Marshak's next letter to the young ecologist, he wrote, "Quite recently, I was asked to communicate to a committee of political scientists any important current literature on the borderlines of their discipline. Can you give me a list of the publications connected with your work on the community of rats?" There is no record of Calhoun's reply, but he must surely have obliged Marshak as he had earlier.[6]

5

MARKED INVASIONS

As the Towson enclosure experiment wound down, and as the role of social behavior in limiting population growth came into focus, Calhoun began to think that to get an even deeper understanding of the relationship between social behavior and population growth, he would need to plant himself amid animal behaviorists and become part of their research community. Over the years, he'd had his fair share of interactions with such folks, like Allee, but he had no formal training in the study of animal behavior, also called ethology.

In 1948 Calhoun read an advertisement in *Science* about a new program in animal behavior at the Roscoe B. Jackson Memorial Laboratory, also known as JAX, in Bar Harbor, Maine. Initially opened in 1929 to study cancer biology, JAX received a five-year Rockefeller Foundation grant in 1945 to initiate a program in animal behavior, with an emphasis on "the relationship between heredity and intelligence, emotions, and other behavior traits in mammals." To lead that program, JAX hired well-known animal behaviorist John Paul Scott, who, among other things, set up a program looking at behavior in man's best friend, the dog. Scott had done his PhD at the University of Chicago, where, like Calhoun, he had his views on behavior and ecology shaped by Allee.

Calhoun visited JAX in the summer of 1948 and was impressed

by both Scott and the facilities—much of the cancer work there focused on mice, and so there were abundant resources to study rodents. Calhoun knew that this was a place that researchers who studied both animal and human behavior were wont to visit, as they had September 10–13, 1946, when JAX hosted a conference on "Genetics and Social Behavior." Given his burgeoning interdisciplinary interests, Calhoun was delighted that the conference flyer outlined, in grandiose language, the problems to be addressed: "The study of such immense and complicated problems as war and peace . . . , and even dominance and territoriality, is not a task for one individual or one laboratory, nor even one field of science. . . . The Conference on 'Genetics and Social Behavior' was called with the idea of collecting a sample of the best minds . . . and getting their advice as to the best methods of attacking the problems cited above."

JAX seemed like the perfect next stop when the enclosure experiment was completed. The problem was there were no funds to support a position for Calhoun there. That changed in April 1949, two months before the enclosure experiment came to a close, when the National Institute of Mental Health (NIMH) created a fellowship that Calhoun learned of from Edith, who was working in the Baltimore County Mental Health Clinic in Towson and had brought home a notice about the fellowship that had come in the clinic's mail that day. Calhoun applied straightaway and was awarded a two-year fellowship to work at JAX. "They [NIMH] have become quite interested in the concept of manipulating the environment to alter behavior patterns," Calhoun wrote his new colleague, economist Jacob Marshak, "and have given me this opportunity to continue my investigation as well at the same time, organizing plans for a possible, long-term study to be sponsored by the National Institute of Mental Health."

Never one to miss an opportunity to talk shop on rodents and population dynamics, Calhoun, on the way from Towson to Bar Harbor, took a detour to the University of Michigan, where in addition to giving a lecture he spent time looking in on a long-term experiment on white-footed mice being run there. Then it was off to Bar Harbor, where he and Edith settled in at their new home, a guesthouse at Eagle Cliff, the country estate of the Luquers, a New York family that traces its Gotham lineage back to the old-money Pierponts. Not long after that, the Calhouns became three, as they welcomed a daughter, Catherine.

It didn't take long before Calhoun realized he had found an ideal academic home to become better versed in the field of animal behavior. He quickly became part of J. P. Scott's Committee for the Study of Animal Societies under Natural Conditions. In the winter of 1949, that committee ran a symposium on "Animal Biology and Sociobiology" at a joint meeting of the American Society of Zoologists and the American Association for the Advancement for Science being held at the Statler Hotel in New York City. The symposium, which took place before a crowd of more than three hundred, had fourteen talks, including presentations by Calhoun (who spoke on the "Influence of Space and Time on the Social Behavior of the Rat"), A. M. Guhl (a collaborator of Allee's who spoke about his work on dominance in chickens), researchers from Scott's dog behavior group, scientists from the New York Zoological Society discussing their work on elephant behavior, and a team from the American Museum of Natural History lecturing on the sex organs of fish. In addition to attending Scott's symposium, Calhoun and others hopped onto the subway to the American Museum of Natural History, where animal behaviorist T. C. Schneirla was hosting an informal think tank on "Problems in the Interpretation of Evidence on Animal Behavior." Between the symposium and the think tank, it was a weeklong dream come true for people who set out to embed themselves into the world of ethology.

Calhoun not only took part in JAX-organized animal behavior symposia like the one in New York—he was soon thinking about organizing a meeting in Bar Harbor. One summer evening, Scott told Calhoun he was working on hosting some sort of symposium at JAX during 1950, but he had not settled on a theme. In short time, Calhoun proposed one. What Calhoun envisioned was tying his interests in animal behavior to issues of human welfare, particularly as it related to problems associated with population growth.[1]

The symposium, which Calhoun tentatively titled "Basic Research into Problems of Behavior and Its Application to Problems of Human Welfare," had a proposed budget of $35,000, ideally to be covered in part by JAX but also by the Carnegie Foundation for Peace and NIMH. Calhoun mapped out panels, including one on behavior in early life, to be chaired by Allee, another on ecology, including ecology in "modern urban man," as well as a panel on the "Management of Man through the Provision of his Basic Needs" that would include talks by Frank Lloyd Wright and renowned city planner Lewis Mumford. Scott considered Calhoun's grandiose proposal, and when all was said and done, culled it down to just one of the questions proposed—how behavior in very early life affected behavior later in life. In 1951, JAX hosted that symposium, "The Effects of Early Experience on Mental Health."

Even when it wasn't hosting symposia, JAX was a magnet for leaders in the field, including both biologists and psychologists interested in mechanisms of animal behavior. In the latter camp, B. F. Skinner visited to give a seminar in the fall of 1949, where Calhoun met him for the first and only time. Skinner was already famous for his work on instrumental (also called operant) learning, in which the behavior of an animal is either rewarded, and so reinforced, or punished, and thereby suppressed. He revolutionized the study of learning with what became known as the Skinner box. The idea was to create a continuous measure of behavior that could somehow be divided into meaningful units. When a rat pushes down on a lever, it is making an *operant response* because the action changes the rat's environment by, for example, adding food to it. Because "lever pushing" is a relatively unambiguous event that is easily measurable, and because it occurs in an environment over which the rat has control, the Skinner box changed the way psychologists studied learning.

Calhoun was no fan of studying animal behavior by testing individual animals in small cages, one at a time, removed from the social milieu in which they had evolved, but he listened carefully as Skinner gave his JAX lecture on behaviorism and operant conditioning. Calhoun and Skinner met after that lecture, but Calhoun doesn't record whether they discussed Skinner's talk or whether they talked of Skinner's recently published (1948) science fiction book, *Walden Two*, in which Frazier, one of the protagonists, built a utopian society by applying Skinner's ideas on behaviorism and manipulating rewards and punishment, but given the two men's interests, it is not implausible that they did.

Calhoun had come to JAX not only to surround himself in an animal behavior milieu but also to continue his research on rodent behavior and population growth. He did that on two parallel fronts: one out in the field and one in the laboratory. The field work took place in the forests of Mount Desert Island, five miles due west of Bar Harbor. As in his work a decade earlier trapping small mammals around Reelfoot Lake, Calhoun organized a large-scale sampling of mice, shrews, and voles on Mount Desert Island. But this time he layered on his experience in population growth and decline from both the streets of Baltimore and the enclosure in Towson. The overarching question for the Mount Desert Island work was this: If population sizes of small mammals in one area were to decline steeply, freeing up space for immigrants, would other small mammals from adjacent areas swarm in to fill the vacuum created? And, if they did, what would happen next?

To find out, Calhoun and his student set up a rectangular grid—bearing a striking resemblance to a city block—of sixty traps, with traps set in clusters of three, separated by fifty feet. The traps were quite effective, capturing about a quarter of the small mammals in the targeted area on day 1 and a quarter of what remained on day 2 and day 3. What Calhoun found was that, in short time, there was a "marked invasion" of small mammals from nearby communities. In fact, by the third week of the experiment, the number of invaders was five times the original population size of the targeted community. Indirect evidence also suggested that when the vacuum was filled (and then some), a chain reaction was indeed set in motion: the vacuum created when the invaders left *their* homes was filled by small mammals near there, and that vacuum was filled by those near there and on and on.

There remained the question of what behavioral mechanisms were in play in the small mammals in the forests of Mount Desert Island. One possibility, Calhoun suggested, was that animals from adjacent areas encountered one another often when patrolling their territories, and the aggressive interactions that occurred when they met provided them behavioral feedback about the size of nearby populations. When the number of aggressive interactions fell, they swarmed in to fill the perceived vacuum. Calhoun's other idea was that animals grew accustomed to a certain level of interactions, both positive and negative, from those around them: when the number of interactions was decreased by the creation of the vacuum, the mice, shrews, and voles moved in the direction that was most likely to increase the interaction level to what it had been before the vacuum.

Calhoun didn't collect behavioral data in the forest, and so he couldn't distinguish between these two hypotheses, but what he did learn was that complex, subtle, and often indirect factors affected population dynamics. He had a sense of that complexity from the experiments he had done with Baltimore street rats, but now he knew that what was true for those city rats was also true, perhaps even more so, for mice, shrews, and voles in the wild—all of which suggested to him that what he learned from small mammals in the forests of Mount Desert Island promised "to be an important key to understanding certain characteristics of the social structure of populations."

On the last day of the experiment, as his student was in the field, Calhoun found himself scanning an army behavior journal that had arrived at the field station library. As he looked through the table of contents, he came across an article by Steve Ranson, who had been Edith's supervisor when they had lived in Towson. That paper was all about mass spatial panic in armed conflicts. As Calhoun worked his way through Ranson's article, it struck him that the mass spatial panic and mass hysterical reactions that happen in war bore some resemblance to the large-scale, chain-reaction movements of the small mammals at Mount Desert Island. Comparing the two, Calhoun thought, might shed light on the "the biology of groups, such as nutritional needs, climatic conditions, interactions of members of a population and interactions among populations."[2]

Mount Desert Island was also where Calhoun did his laboratory work, as the island was home to JAX's Hamilton field station, a gift of the late William Pierson Hamilton, one of J. P. Morgan's partners. The rodent facilities were in the basement of the main building, and because of JAX's focus on cancer, the space was designed to hold hundreds of house mice but no rats. And so, Calhoun switched his experimental study species but kept his research focused on behavior and population dynamics, except now he was also coming at these matters informed not only by ecology and animal behavior but also by new ideas from the literature on human behavior and population structure.

"No sooner had I arrived in Bar Harbor," Calhoun wrote, "that a chance encounter led me to John Q. Stewart's article on 'Social Physics.'" Stewart, an astrophysicist at Princeton, had published that article in *Scientific American* a year earlier. In Stewart's paper, "Concerning 'Social Physics,'" he opens by telling his readers, including Calhoun, that the quotation marks around "social physics" in the title of the article "indicate that it is not an accepted science, although it may well become one. Its principal concept: the behavior of people in large numbers may be predicted by mathematical rules."

Social physics—using the mathematics of physics to predict social behavior—had originated with French philosopher and mathematician Auguste Comte. To get the core ideas of the still nascent subfield across to readers of *Scientific American*, Stewart asked them to image a giant who was so large that the earth appeared the size of a molecule to him: "From such a vantage point," Stewart wrote, "an observer equipped with appropriate devices for measuring population densities and movements might discover that, like molecules in a gas, groups of men obey certain simple physical laws." Those laws were built on statistical measures of space, time, and number in humans. Stewart granted that people are more complex and unpredictable than atoms and molecules but then reminded readers that Heisenberg's uncertainty principle shows atoms are also an unpredictable lot, yet there are mathematical ways to handle their behavior that can be applied to humans, and that was precisely what social physicists did.

As a case in point, Stewart turned to the studies of Harvard linguist and philosopher George Zipf, who had done work where he found that to gauge the relative number of telephone calls between people in any two cities in the United States, you multiply the size of the cities and then divide that by the distance between them: the same approach, Zipf showed, could be used for all sorts of things besides the number of phone calls (e.g., the number of bus tickets sold between city A and city B). Working off Zipf's findings, Stewart argued that one can take something akin to Newton's law of universal gravitation and transform it to the language of social physics to explain the sorts of phenomena Zipf discussed: in lieu of Newton's particles in the universe and the distance between those particles, social physics used the number of people in different populations and the distance between those populations.

Calhoun was struck by what Stewart and Zipf were proposing. In time he would adopt the language of social physics: "The basic particle is the individual mammal. In any total assembly of such particles inhabiting a particular environment, taxonomic categories, such as species and genera, represent general classes of particles." But in 1949, his thoughts were more practical: the idea "of every individual being affected by every other individual, inversely proportional to the intervening distance between them," opened a door of understanding "to the dynamics of a complex social system," including the social system of animals, like mice. As he began mapping out how to test these ideas in his lab at the Hamilton Station, Calhoun read Zipf's hot-off-the-press book *Human Behavior and the Principle of Least Effort*, and he began a long-term correspondence, and then friendship, with both Stewart and Zipf.

Using two strains of mice that had been bred for hundreds of generations at Hamilton Station, Calhoun designed a pair of experiments that tied together his ecological and ethological interests in populations to burgeoning ideas in social physics. He also added in a medical slant, since so much work at the Jackson Labs was on cancer. Mice in both strains were of similar size but differed in what Calhoun called "physiological stability." Mice in the physiologically unstable strain had high incidences of breast cancer and displayed seizures in response to certain loud sounds, while mice in the physiologically stable strain showed neither malady.

Calhoun's lab was about four hundred square feet, which he divided into four sections to allow multiple trials to go on simultaneously. Not much was known at the time about home ranges in wild mice and the few estimates available ranged from four thousand to forty thousand square feet. Even taking that lower estimate as a starting point, it was clear to Calhoun that if one of the things he was going to study was distance as it related to behavior in populations, the floor space he had to work with was far too small.

Like an urban planner of his day, Calhoun reasoned that if he—and more importantly the mice—couldn't spread out, he'd have to build up. On the floor of the experimental pen sat two dining halls with a total of three food hoppers and two water stations. There were also two boxes filled with nesting material that mice could draw from at will. In most trials in the experiments, the mice lived in two rodent apartment buildings that Calhoun constructed. Each building had four floors, separated by eighteen inches, and each floor was accessible by its own ramp set at a sixty-degree angle between the apartment floor and the bottom of the pen. Each floor of a building had what Calhoun called "places of abode"—nesting box apartments in which mice lived. Calhoun could watch what was happening from an observation booth, and when he couldn't be watching, he could infer basic movement patterns by following the trail of mouse feces.

Building up created more space, but not the ten to a hundred times more space that would be needed to mimic a true home range in the wild. But, Calhoun reasoned, it would take far more energy to scale the vertical space he created than to walk on the floor, and that needed to be taken into account. He found studies showing that people expend fifteen times more energy walking up a flight of stairs and consume seventeen times the oxygen, as compared to walking the same distance on a flat surface: "Since it is logical to assume that mice also expend more energy in traversing inclined routes of travel," he wrote, "this vertical structuring of the environment must have effectively expanded the available space."

Mouse residents were certainly considering the distance from their apartment to the dining hall and boxes filled with nesting material. Though they explored the top floors of apartment buildings, they tended to move into apartments on the lower floors. They much preferred time in those apartments to time on the pen floor, where they rarely stopped except to eat, drink, or grab some nesting material, though motivation mattered, as hungry mice took the shortest, fastest route to the dining hall and nesting material, while more satiated mice dawdled some.

The rats of Baltimore had taught Calhoun that aggression plays an important role in population dynamics, and so he kept a keen watch on that in his mice. In both strains, there was a great deal of variability in how likely a mouse was to be in areas where battles for status occurred: some mice preferred such areas; others actively avoided them. Between-strain differences also emerged. Mice in the physiologically unstable strain displayed more aggression and fewer instances of prosocial behaviors, such as grooming, than mice in the physiologically stable strain: "a modicum of mild aggression in . . . a well-organized society" was how Calhoun summarized life for mice from the physiologically stable strain, but for those in the unstable strain, "this selfsame environment became a hellhole source of continuing tension."

Calhoun also noted other important behavioral differences between the two strains. Physiologically unstable mice tended to show more stereotypical "hardwired" behaviors, while it seemed to Calhoun that mice in the physiologically stable strain were better learners, adjusting to changing conditions sooner and more efficiently than their counterparts. Calhoun also found differences in maternal behavior, with mothers in the physiologically unstable strain displaying less parental care than physiologically stable mothers. All in all, mice in the physiologically unstable strain were, in a sense, behaviorally unstable, and mice in the physiologically stable strain were behaviorally stable.

One thing that struck Calhoun was that although hereditary differences between the strains surely contributed to the behavioral differences, they weren't the whole story. Population size also mattered. As population size increased, the behavioral differences between the physiologically unstable and the physiologically stable strains began to shrink. It wasn't so much that physiologically unstable mice did anything different as it was that physiologically stable males became more aggressive and physiologically stable mothers became poor providers: increasing population size, Calhoun came to think, had turned behaviorally stable mice into a much more unstable lot.[3]

As he studied population dynamics and behavior in more species, Calhoun read a plethora of papers and books from the sociological, psychological, and psychiatric literature and became more of an integrative thinker, searching for pattern and process as they relate to population dynamics and behavior. When he wasn't in the field, in the laboratory, or reading, Calhoun was writing and lecturing whenever and wherever the opportunity arose. He published review papers such as "The Study of Wild Animals under Controlled Conditions" and "The Social Aspects of Population Dynamics." In the latter paper, Calhoun relied heavily on the work of Allee and others at the University of Chicago, as well as J. P. Scott at JAX.

Though "The Social Aspects of Population Dynamics" is a review paper largely about behavior and population dynamics in animals, there are hints of Calhoun's interests in human population dynamics.

He notes an inherent conflict between two forces—increasing population growth and an individual attaining its goals—causing an increase in social tension. "As our human society becomes more highly technical," Calhoun wrote, "there arises the tendency to accentuate one of these growth phenomena without any consideration as to its effect on other aspects of life. This is particularly so as regards increasing population density, which in many quarters is accepted as a desirable objective." That desirable objective, Calhoun thought, was anything but. For Calhoun, evolutionary biologist William Bateson had summed it up nicely: "To spread a layer of human protoplasm of the greatest possible thickness over the earth—the implied ambition of many publicists—in the light of natural knowledge is seen to be reckless folly."

In February 1950, Calhoun presented a lecture in which he took his boldest, arguably most outlandish, initial stabs at breaking down the boundaries that separated ecology, animal behavior, human sociology, and human psychology. He presented that lecture, "Ecological Principles and the Concept of Morality," to his colleagues at Hamilton Station, and after railing against specialization in the sciences—what Calhoun called the "schismatism" of the sciences—he told his audience that he was there to argue that the birthright of the field of ecology was the study of the complex web of interactions that make up life on earth. To do this, Calhoun continued, he needed "a common denominator," something that applies to all life, and for that he selected an organism's ability to utilize energy.

What that all boiled down to was, as soon as life emerged, evolution worked to produce lifeforms that were efficient at storing solar energy. That maximization first applied to the growth of individuals and then, over longer evolutionary stretches, the growth of communities made up of many species. Those communities were where what Calhoun called "unconscious morality" came into play. "The greater the diversity," he told his colleagues, "the less the effect will there be on the whole community of a deleterious action against a single component unit": diversity buffers all members of a community against harm that befalls one species in its midst. But that greater diversity and interconnectivity came at a cost, because "the greater will be the effect of a slight change which affects all members," which is to say that ecological food webs (as well as other webs) in such communities were so complex that changes in any part would snowball and force changes to the other parts.

Today, Catherine (Cat) Calhoun recalls her father as a man who "didn't think in a linear fashion . . . [his] mind would go here," she says, "and then it would go over there. . . . People would just stand there like, 'What the hell is he talking about?'" John Calhoun's lecture on "Ecological Principles and the Concept of Morality" must have struck listeners in a similar vein: as a rambling, stream-of-consciousness discourse that at times was hard to make heads or tails out of, likely because Calhoun was still sorting this out himself, not at all clear as to what intellectual destination his various trains of thoughts might lead him. And as his two-year fellowship at JAX approached its completion in June 1951, he also didn't know what the next geographical destination on his journey as a scientist would be.

Angling for some sort of position at NIMH, right before his JAX fellowship ended, Calhoun took a calculated risk and he and his family headed to Washington, DC. He heard NIMH was attempting to acquire or lease a large barn for experimental work on animals, which he had heard rumors of when the NIH scientists had visited him at Towson. Calhoun was convinced this would be the perfect home for his work on rodent population dynamics and growth.

On the drive to Washington, Calhoun took a detour to the Worcester Foundation for Experimental Biology at Clark University in Massachusetts, to lobby its director, Dr. Hudson Hoagland, who sat on a panel involved in funding decisions at NIMH. Hoagland promised to raise Calhoun's work to the NIMH panel, which he did, but nothing more came of it. When Calhoun arrived in Washington, he took matters into his own hands and met with some lower-level administrative people at NIMH, but nothing "sufficed to move my case to a hiring stage." Still unemployed, Calhoun next reached out to Steve

Ranson, Edith's former boss, who by then was part of the office of the Army Surgeon General.

After reading Ranson's paper on mass panic in the army journal about a year earlier, Calhoun had reestablished a connection. When Calhoun inquired whether Ranson might have any leads that he could follow as the end of his fellowship approached, Ranson told him of a newly formed behavioral research team at the Walter Reed Army Medical Center in Silver Spring, Maryland, about an hour's drive southwest of where the Calhouns had lived in Towson. Ranson called Dr. David Rioch, a psychiatrist with an interest in the brain and neuroanatomy, who was the director of that new research group, and told him a bit about Calhoun and his work. Rioch was interested enough to invite Calhoun for an interview. "On arriving at Walter Reed," Calhoun recalled, "I found Dr. Rioch . . . in a large room sparsely furnished with old furniture. I was welcomed like a long lost child and we spent many hours together during the next few days, [with] Dr. Rioch . . . listening to my ideas and hopes." Rioch wanted Calhoun as part of his team at Walter Reed, but there was a problem. Rioch told Calhoun that "he could never get [Calhoun] on the staff as an ecologist." Calhoun could sympathize: "Who in the medical establishment had ever heard of an ecologist!" he wrote, "So he made a 'research psychologist' out of me."

Six weeks later, Calhoun and his family had moved to Silver Spring, and for the next four years, his work address would read John Calhoun, PhD, Research Psychologist, Neuropsychiatry Division, Army Medical Service Graduate School, Walter Reed Army Medical Center.[4]

6

AN INKLING INTO HOW PANIC MIGHT BE INDUCED

David Rioch's group in the Neuropsychiatry Division of the Army Medical Service Graduate School at Walter Reed was a good, but not a perfect, fit for Calhoun's burgeoning, cross-disciplinary interests in behavior and population dynamics. Rioch's team was the result of the army wanting to deal with what they called "psychiatric casualties." The emphasis on a topic that he knew nothing about didn't concern Calhoun. From their conversations, Rioch knew very well what Calhoun's interests and expertise were, and he was excited to have him on board. Calhoun was more concerned that many of the other new members in Rioch's team were "Skinnerian grandchildren," who, Calhoun recalled, "filled rooms with Skinner boxes and cumulative recorders . . . [and] reinforced the unease I felt on first listening to Skinner himself at Bar Harbor." Still, Skinner's disciples would be cloistered in their own lab space, and so the problem there was more philosophical than practical. What's more, Rioch's group at Walter Reed drew visits from a bevy of interesting scholars. One was anthropologist Edward Hall, who in time would become famous for his work on "proxemics," which he defined as "the interrelated observations and theories of man's use of space." Rioch "encouraged [my] intellectual collaboration [with Hall]," Calhoun recalled, but given the two men's mutual interest in the use of space, that encouragement was hardly necessary, and Calhoun and Hall soon became both colleagues and friends.

One thing the position at Walter Reed gave Calhoun was the time and freedom to analyze the reams of data he had collected during the Towson enclosure experiment. He estimated that half his time in Rioch's lab was spent taking all that data and writing it up as a monograph, *The Ecology and Sociology of the Norway Rat*. But, as he had at so many of his research homes, Calhoun also had one foot in the field and the other foot in the laboratory.

Rioch had sold Calhoun's hire to the army brass based on the work he had done on chain-reaction movements in mammals. "The Korean War was then in full swing, and the ostensible justification," Calhoun wrote, "was that [my work] might provide some inkling into how panic might be induced in enemy troops." Calhoun and Rioch put together a small contract, which the army funded, to expand on the chain-reaction movements in mammals by repeating the field work that Calhoun had done on Mount Desert Island in Maine, but this time he would work in the Huntington Wildlife Forest near Newcomb, New York.

The on-the-ground study at Huntington was led by William Webb, a professor in the College of Forestry at the State University of New York in Syracuse, who worked closely with Calhoun. Webb, like Calhoun, found that when small mammals from a community were trapped, that vacuum was filled by others from nearby communities, setting into motion chain-reaction movements that filled each new vacuum as it opened. If, Calhoun told David Rioch, panic was defined as "the marked repetition of a single behavior, or small set of behaviors to the point that insufficient or inappropriate responses are made," then neither the work at Bar Harbor nor New York was evidence of panic, let alone mass panic. But Calhoun added, "if progressively accentuated," what he was seeing would "terminate in . . . a state of panic."

Calhoun and Webb learned more about the behavioral dominance hierarchy *between* small-mammal species in the New York communities than Calhoun had at Mount Desert Island, and their results added a new layer to this tale of chain reactions: which species moved into a vacuum depended on which species had been vacated to create the vacuum. *If* the species removed were behaviorally dominant, the void that their removal created was likely to be filled by individuals from behaviorally subordinate species: individuals who stood little chance to forcibly evict others were quick to sense a land grab opportunity and just as quick to take advantage of it.[1]

From the start, Calhoun saw Walter Reed as a stopover until he could find a way to land a position at nearby NIMH. He was confident enough this might happen that, with another daughter (Cheshire) on the way, he and Edith purchased a home in Kensington, three miles northeast of both Walter Reed and NIMH. By late 1953, Calhoun became concerned that the move to NIMH needed to happen sooner rather than later, as his funding at Walter Reed was becoming more tenuous, largely because of one man—Republican senator Joe McCarthy.

McCarthy's delusions included not just that communists had infiltrated every nook and cranny of the government but, more generally, that politicians from across the aisle were trying to undermine the armed services by forwarding their liberal agenda. The colonel who administered Walter Reed, "began to worry what McCarthy might think," Calhoun wrote, "if he found out that the Army was spending money studying the social behavior of mice and shrews in the New York mountains." As Calhoun saw it, in the colonel's eyes "sociology = socialism = communism: the U.S. Army must be moribund supporting communism," which meant it was just a matter of time before he would get his "walking papers."

To prepare for what he saw as the inevitable, Calhoun contacted David Shakow, an experimental psychologist, who was the head of a new Laboratory of Psychology at NIMH. "The area of research which I propose is that of social behavior and social organization of animals," Calhoun wrote Shakow, campaigning for a position. "The origin of

[my] theoretical framework is dependent primarily upon my formal training and experience in ecology and secondarily by the close association I've had with psychologists and psychiatrists during the past seven years." To justify leaving Walter Reed, Calhoun told Shakow of the "curtailment of funds" for his work there: Calhoun later annotated his copy of that letter, circling "curtailment of funds" and adding "a nice way of phrasing that the Colonel in charge of [Rioch's group] was afraid of Joe McCarthy."

In the pages that followed his cover letter, Calhoun details what he'd focus on if Shakow could arrange a position for him. "My interest in group dynamics on the infra-human level," wrote Calhoun, "is directed toward the elucidation of those principles, which govern the density of a population, and the stress experienced by its members." To study those principles, he outlined a series of experiments studying home range, social interactions, formation of groups, stress, and more.

In the first draft of his proposal to Shakow, Calhoun wrote, but then opted to omit, that his work would "in a sense . . . form a 'preventive medicine' of social behavior . . . those principles . . . which prevent the formation of social instability of groups, and social malfunction of the individual is of equal, if not greater importance, than those principles that govern the correction of social malfunction." Recognizing that what he proposed was not the sort of research currently underway in Shakow's laboratory, Calhoun suggested a panel of animal behaviorists to assess the progress of his research program.

As this was unfolding, Calhoun would occasionally get together and talk shop with his friend Robert Cook, who was director at the Population Reference Bureau, a private foundation that gathers population data from across the globe. One day, when Calhoun was "at the end of all [his] hopes for continuing in population research," Cook told him that Alan Gregg, who had been heading a major branch of the Rockefeller Foundation, and who had many connections at NIMH, would be in town. Cook told Calhoun he'd arrange lunch for the three of them at the Cosmos Club in Washington. Not long after that lunch, likely because of a few strategic phone calls by Gregg, Shakow contacted Calhoun. "Exactly what transpired, I will never know," Calhoun wrote, but a behind-closed-doors agreement had been reached in which Calhoun would join the Perception and Learning section of Shakow's Laboratory of Psychology, on the condition that the army, via Walter Reed, paid his first year's salary ($8,990) at NIMH, which they agreed to do.

To make all this official, NIMH had to advertise a position for a "Psychologist (Physiological and Experimental)." Shakow asked Calhoun to craft that ad, which was easy enough for Calhoun to do: the first few sentences called for a "senior investigator in the comparative (phyletic) studies of group behavior in animals" who had experience in "a variety of sciences including psychology, ecology . . . and physiology," who could "design and conduct research . . . [on] the relative stability of populations; the manner in which stress and tension arise from particular patterns of interactions[;] . . . how interaction patterns are determined by the physical and social environment . . . [and] obtain data regarding the status of natural populations of mammals." Though "no applicants besides John Calhoun need bother apply" was not part of the advertisement, it might as well have been.[2]

In August of 1954, when Calhoun moved from Walter Reed to the Perception and Learning section of the Laboratory of Psychology, NIMH had only recently been granted access to an experimental barn like the one that rumors had swirled about for years. Two months prior to Calhoun's arrival, NIMH had worked out a deal with Eugene Casey, whereby Casey granted the agency a six-year lease—free of charge—to a huge dairy barn and two attached smaller barns that were part of his farm in Rockville, Maryland. Not long after that, Casey agreed to cover the costs of the major reconstruction work that would be needed to turn the second floor of the barn into an area where experiments could be conducted: the first floor would remain a working barn for Casey, with cows and more residing there. In return, Casey asked only that he be apprised of the projects undertaken and "what, if any, contributions to medical or other sciences ha[ve] been obtained from research directly related to our contract."

During his first year, with his salary covered by the army, Calhoun didn't have much clout at NIMH, but he saw the Casey Barn as an eventual home for his future work on rats, behavior, and population dynamics, and so as soon as he was drawing a salary directly from NIMH, he began lobbying for space. Not only did he have the full support of the Laboratory of Psychology but he also had support from those much higher up in the NIMH bureaucracy, including the director of NIH and the director of NIMH.

No one else was campaigning for the space, so it wasn't long before Calhoun was granted a large part of the second floor of the barn for his own use. Calhoun thought that this might be the only opportunity he would have to design and oversee the construction of an optimal space to conduct his research. He also knew that to turn that space into what he needed would require a large and expensive remodeling effort, involving not just a complete overhaul of the electrical, plumbing, and ventilation systems but also the building of an office, six experimental rooms (partitioned by new walls), animal and storage facilities, and food storage facilities. It would also mean raising certain parts of the floor and building a subroof with a three-by-five-foot glass panel strategically placed to allow Calhoun or one of his team to lie down and observe any experiments under way. Given the pace at which contractors, particularly those working for the federal government operated, Calhoun put an estimate of about two years to get all the work done. He couldn't start any experiments until then, so he immediately took to micromanaging the construction work, as well as making the case to NIMH for funds to pay for the personnel, including an animal caretaker and a secretary, that Calhoun would eventually need to manage his studies in the barn.[3]

In a series of blueprints and master plans that he was constantly revising and updating throughout 1955–56, Calhoun laid out how the experimental rooms in the barn should be designed to allow him the flexibility to do any number of different experiments on behavior and population dynamics in rats. Three rooms would each house ten-by-fourteen-foot enclosures. What experiments would take place in the enclosures were to be determined, but Calhoun knew that he would employ them to investigate "complex social group[s] having reproductive continuity through several generations."

Inspired in part by Allee's work on social behavior, another area in Calhoun's lab-to-be was designed for experimental work on how cooperation, or its converse—what Calhoun, like Allee before him, called disoperation—affected behavioral dynamics and population growth in rats. Calhoun requested that a custom-built device he called a STAW—socialization training apparatus for water—be constructed for this experiment. The STAW had two levers attached to a water bottle. Though rats would be rewarded for pushing a lever, this was no Skinnerian operant conditioning experiment in which a lone rat's existence revolves around a lever for a reward. Instead, rats in Calhoun's STAW device were living in a large open environment and embedded in social groups (as large as thirty-two), experiencing all the complex interactions that occur in such groups. What's more, pressing on a lever only provided water under certain socially dictated conditions.

In the cooperation treatment of Calhoun's experiments, two rats, separated by a clear divider, would both need to hold down a lever for either to get water; in the disoperation treatment, if one rat was holding down a lever and a second rat came over and held down the other lever, the water stopped flowing. In a letter to Eugene Casey, who had leased the barn to NIMH, Calhoun described the cooperation treatment as akin to "living in a small town where most every relation one has with his family and neighbors is a happy and satisfactory one" and the disoperation being analogous to "living in a small town where a host of conditions makes one continually suspicious of the intent of his neighbors right through from birth to death."

In the cooperation/disoperation experiment, Calhoun decided he would use rats, living in adjacent but disconnected pens. In one condition, pens with the STAW device would be set to cooperation, and in the other to disoperation. This was meant to be a long-term experiment, spanning multiple generations in each pen, with Calhoun fantasizing that it might go on, undisturbed, "for a period of at least five years in order to permit maturation of the populations [and] development of culture." The overarching aim of the experiment would be not just to test whether rats could learn to jointly press down levers or avoid doing so—based on the literature on rat learning, Calhoun was fairly certain they would—but rather to examine how cooperation and disoperation affected sociality in groups and population growth over generations.[4]

As he oversaw the barn remodel, Calhoun kept himself busy on several other fronts. Leonard Duhl, a psychiatrist at NIMH, who would go on to become known as the grandfather of the Healthy Cities initiative adopted by the World Health Organization, was building an NIMH-funded think tank called the Committee to Explore the Influence of Physical Social Environmental Variables as Determinants of Mental Health, but which members quickly dubbed the Space Cadets because of the group's interest in the effect of physical space on behavior. With his embryonic ideas on the relationship between population growth in animals and humans starting to mature, Calhoun was happy to join the Space Cadets and soon thereafter share the leadership role with Duhl.

The Space Cadets were an interdisciplinary lot that included Calhoun's colleagues, animal behaviorist T. C. Schneirla and social physicist John Stewart, as well as the University of Chicago mathematical biologist and physicist Nicolas Rashevsky, landscape architect/ecologist Ian McHarg, city planners, economists, anthropologists, medical sociologists, and more. The Space Cadets would hold informal meetings for two to three days each year for more than a decade. Largely a think tank to exchange ideas and mutually enlighten one another, the Space Cadets was a group where members from very different disciplines would, on occasion, collaborate. Calhoun was at the head of that list of those engaged in interdisciplinary collaborations.

After participating in a 1955 meeting of the American Association for the Advancement of Science's Committee on Social Physics, Calhoun began working on a more formal mathematical way to capture his thoughts on social behavior and population growth. Though fairly well-versed in statistics, he didn't have the mathematical sophistication to do that alone. To remedy that, he recruited collaborators: Murray Eden, an engineer and biophysicist who was a visiting fellow at the National Heart Institute, and fellow Space Cadet Rashevsky, the author of a number of books on biology and mathematics.

Calhoun was beginning to think the literature on population growth in both animals and humans was too myopic: "The concept of carrying capacity," he wrote of the population size at which population growth levels off, "implies maximizing numbers alone." But Calhoun thought there was much more to be maximized than just numbers of individuals. What he dubbed "social welfare" was concerned not just with numbers and reproductive rates but with what would maximize the nutritional status of the individuals in a population and their life expectancy, "buffering the individual from the vicissitudes of his physical environment" and fostering "the development of stable and mutually satisfactory social relationships."

The mathematical model that Calhoun, Eden, and Rashevsky developed was complex, but what was most important to Calhoun was not the endless equations involved, or even some of the model's predictions (which were still somewhat vague), but rather that his model of social welfare revolved around just three variables: the number of animals interacting with one another, the time between when an animal ended one social interaction and when it began another (the "refractory" period), and the size of the space in which social interactions took place. Each of these variables was one that Calhoun had worked with and that he planned to dig deeper into once he began experiments in the Casey Barn, and so he found it reassuring that, in principle, they could be used to model how social behavior affected population dynamics.

Calhoun first presented his ideas on social welfare as part of a 1957 Cold Spring Harbor Laboratory Symposium on "Population Studies:

The Casey Barn in the early 1940s, when it was used as a dairy barn and its outer walls were employed as sign boards. Calhoun did not move into the barn for more than a decade and a half after this, but the structure of the barn did not change significantly over those years. Photo courtesy of GaithersburgHistory.com.

Animal Ecology and Demography." He knew that an invitation to speak at a Cold Spring Harbor Symposium was a badge of honor; only four years earlier, James Watson had given the first lecture on his and Francis Crick's pathbreaking discovery of the double-helix structure of DNA as part of a Cold Spring Harbor Symposium on viruses. And though Watson and Crick were not at the 1957 symposium to hear Calhoun lecture on "Social Welfare as a Variable in Population Dynamics," other luminaries—such as Ernst Mayr and Theodosius Dobzhansky, two key players in what became known as the evolutionary synthesis, which brought evolutionary biology, population genetics, paleontology, and more together under one umbrella—were present, as was G. Evelyn Hutchinson, arguably the most important ecologist of the period. They, and many of the other 150 scientists present that day, heard Calhoun speak about his three critical social welfare variables, explain what they looked like in a mathematical model, and present a very brief discussion of his own work "on the utilization of time and space by rats," particularly the role age and prior experiences played in the refractory period of the rats he worked with on the streets of Baltimore and in his Towson enclosure.

When Calhoun returned from Cold Spring Harbor in mid-June 1957, he could see that though construction in the barn was coming along, his estimate of two years (spring or summer 1957) had been conservative, as issues regarding potential fire hazards, as well as just the complexity of an operation involving so many working parts, had slowed things to the point that he now estimated a move-in date of very late in the year.

At long last, on December 2, 1957, Calhoun moved into the renovated experimental barn with an understanding from Casey and NIMH that he would have it to himself for a period of four years.[5]

7

RATS ARE NOT MEN, BUT . . .

As Calhoun began his work in the Casey Barn, new ideas were emerging and reshaping the field of animal behavior. Largely working with birds and fish and using a combination of natural history studies and controlled experiments, Konrad Lorenz, Niko Tinbergen, and Karl von Frisch had demonstrated just how complex animal behavior was and, perhaps more importantly, that this complexity was the product of natural selection.

Von Frisch had done his seminal work on how honeybee workers communicate information about any food they find to other members of their hive to recruit more workers to the newly discovered bounty. The answer, von Frisch discovered, lay in the exquisite "waggle dance" that a returning worker acts out when she—workers are always female—arrives at the hive. Imagine a bee returning from a cluster of flowers that is located one thousand feet from her nest, and imagine that the flowers are located forty degrees west of an imaginary straight line running between the worker's nest and the sun. When a worker is back inside the hive, she "dances" up and down a vertical honeycomb waggling her abdomen. The waggle dance provides topographical information (north, east, south, west) about where food is in relation to the hive: compared with a straight up-and-down run along a honeybee comb, the angle at which the forager dances—in this case forty degrees—provides information about the position of the food in relation to the hive and to the sun. The longer the worker dances, the farther away the bounty.

In time, von Frisch, Lorenz, and Tinbergen would share a Nobel Prize "for their discoveries concerning organization and elicitation of individual and social behaviour patterns." As the Nobel Committee noted in its press release, these three had shown that "behaviour patterns become explicable when interpreted as the result of natural selection, analogous with anatomical and physiological characteristics," and the trio were no less than "the most eminent founders of a new science, called 'the comparative study of behaviour' or 'ethology' (from ethos = habit, manner)."

The work leading to that Nobel Prize focused on behaviors that had genetic predispositions. Indeed, Tinbergen's 1951 book was titled *The Study of Instinct*. But other work was showing that behavioral complexity in animals could be manifested in a different way, through a rudimentary type of culture. Syunzo Kawamura and his colleagues were studying troops of Japanese macaques on Koshima Islet, Japan, in the 1950s, when one monkey, Imo, did something that eventually made her a celebrity in the world of ethology. Kawamura's team had been throwing sweet potatoes on the sandy beach for the monkeys to gather and eat. When Imo (whose name is Japanese for potato, or tuber) was a year old, she began to wash the sweet potatoes in water before she ate them. Both novel and creative behavior, this act, which was never before seen in Imo's population, allowed Imo to remove all the sand from the sweet potatoes before she ate them. That alone caught Kawamura's eye, but even more remarkable was that many of Imo's friends and relatives learned the skill of potato washing by watching Imo, in what Kawamura called a simple form of culture.

Ethologists were also learning that it wasn't only primates that were capable of this sort of simple form of culture: birds were as well. During the mid-twentieth century, Brits had their milk delivered to their front porches in bottles with foil caps. "Birds described as tits were observed to prise open the wax board tops of milk bottles on the doorstep . . . [and] drink the milk," according to animal behaviorists

James Fisher and Robert Hinde, who described the first occurrences of this innovative way to get a meal. "[This] has now become a widespread habit in many parts of England and some parts of Wales, Scotland, and Ireland," Fisher and Hinde noted. "The bottles are usually attacked within a few minutes of being left at the door." There were even numerous reports of birds tracking a milkman down the street and opening the bottles on his cart when he had stopped to make a delivery. What's more, different grades of milk had different colored foil covers, and birds in an area where milk bottles were opened tended to prefer the same color foils. In 1949, Fisher and Hinde circulated a survey to hundreds of members of the British Ornithological Society about this innovative means of acquiring a protein-packed drink. Based on what they learned from that survey, they proposed that milk-bottle opening was, on occasion, accidentally stumbled upon by a lucky bird and that others learned this nifty trick, at least in part, from watching the original milk thief. Again, a simple form of culture.

Calhoun was well aware of all these advances in the field of animal behavior. Over the last decade he had taken a lead on J. P. Scott's Committee for the Study of Animal Societies under Natural Conditions and was an important member of the Committee on Sociobiology and Animal Behavior, both of which were integral in creating animal behavior sections in the British Ecological Society and the American Zoological Society. In 1953, both committees Calhoun sat on had helped establish the journal *Animal Behavior*. The Animal Behavior Society, which celebrates its sixtieth birthday in 2024, lists Calhoun as one of its founding members.

One thing Calhoun learned from the work on honeybees, Japanese macaques, tits, and many other species was that his rats in the Casey Barn might employ complex, subtle social behaviors that may have slipped under his radar back in Baltimore and Towson.[1]

The study of ecology and population biology was undergoing changes that were just as dramatic as those in the field of animal behavior.

And, again, Calhoun was in the midst of it all. At the Cold Spring Harbor meeting he spoke at in 1957, G. Evelyn Hutchinson outlined his idea of the fundamental niche. Hutchinson proposed that when all environmental variables are mapped onto a multidimensional graph, the fundamental niche "corresponds to a possible environmental state permitting the species to exist indefinitely." In other words, the fundamental niche was a way to measure what sort of environment would allow a species to at least maintain, and perhaps increase, population size.

Calhoun was not as interested in fundamental niches as he was in a debate that was going on about the dynamics of population growth in animals. That debate centered on the question of how animals control population growth. Calhoun thought aggression regulated the population growth of rats on the streets of Baltimore, but it was a more overarching question that was at the heart of the debate on population growth when the work in the Casey Barn was about to commence. Did individuals always seek to optimize their own rate of reproduction, or was it possible that, as their population reached carrying capacity, individuals would temper their reproduction to avoid overexploiting their resources and potentially causing a population crash as a result?

On one side of the debate was David Lack, a Fellow of the Royal Society of London and an ornithologist and population ecologist at the University of Oxford's Edward Grey Institute of Ornithology. Using his own observations in the Galápagos Islands, as well as work from others, Lack made the case that natural selection favored producing the largest clutch size parents could reliably feed. If, at times, it appeared to ecologists that birds produced fewer chicks than they could feed, it was an illusion: instead, they were optimizing their reproduction constrained only by environmental conditions, such as the availability of food. When food was scarce, clutches were small; but when it was abundant, clutches were large. "Clutches above the normal limit are at a disadvantage, because the young are weakened through undernourishment," Lack wrote in an influential 1954 book,

The Natural Regulation of Animal Numbers, "and as a result, fewer survive per brood from clutches of normal size."

V. C. Wynne-Edwards, a Fellow of the Royal Society of Edinburgh and an ornithologist at Aberdeen University, thought otherwise. Based partly on his own work with red grouse, Wynne-Edwards proposed several mechanisms by which individuals regulate their numbers as to avoid their population overexploiting resources and crashing. Territoriality was one such mechanism, as it reduced the number of breeding pairs in a given area and so maintained a population at sustainable levels. But Wynne-Edwards's most celebrated example of population regulation was the daily chorusing behavior heard in many species of birds. Such chorusing, Wynne-Edwards hypothesized, was a mechanism of censusing population density. When such censuses began to hint that the population was nearing carrying capacity, individuals would curtail reproduction to avoid overutilization of resources and population-level extinction. This model of social behavior has been called group selection because, Wynne-Edwards argued, groups with individuals who behaved this way are more likely to survive and thrive than groups whose constituents ignored the census data.

Both Lack's and Wynne-Edwards's hypotheses placed natural selection front and center in governing population regulation. The difference lay in what was being selected: for Lack, it was always individual reproductive success; for Wynne-Edwards, it was population survival. In the long run, most animal behaviorists would come to reject Wynne-Edwards's ideas, but at the time Calhoun was sympathetic to Wynne-Edwards's group-selection models of social behavior, largely because his mentor in animal behavior, Allee, was. On the other hand, he realized that it might very well be that populations of rats in Baltimore never approached their carrying capacity—not because they were curtailing reproduction in an effort to avoid overexploiting resources but rather because subordinate rats had low reproductive success. Subordinates had mating attempts quashed by dominant males and by females unreceptive to those at the bottom of the dominance hierarchy, and so they may have been making the best of a bad job in their subordinate role, biding their time, until perhaps circumstances were more favorable. Calhoun hoped that, among other things, the detailed cross-generational observations he was about to start might help shed light on the Lack versus Wynne-Edwards debate.

It was an exciting time to be studying animal behavior, ecology, and population growth, and Calhoun held great hopes for both the cooperation experiment and the other, still not-quite-developed, experiments he'd run on the second floor of the Casey Barn.[2]

The cooperation experiment got off to a good start. With lots of space to work with, Calhoun had built twenty-four pens, each measuring two and a half by five feet. Each pen had a STAW (socialization training apparatus for water) device installed within it: in half the pens, the device was set so that both rats needed to cooperate and press on a lever at the same time for either to get water, and in half it was set to "disoperate," so that rats could only get water if another rat was not pressing on a lever. He also decided to add another variable into the mix, one not in his original master plan: Calhoun would vary the number of rats in a pen, with some pens being home to a pair of rats and others either four, eight, sixteen, or thirty-two rats.

Calhoun found that it was the rats in groups of eight cooperators who produced the most offspring, but it was the way in which group size affected learning in both the cooperation and disoperation groups that really struck him. In both groups the rats needed to learn something: either that another rat near a lever meant an opportunity to get a drink or that another rat in the vicinity of a lever meant the odds of getting a drink were low. In most cooperation pens, the rats learned to "cooperate with marked precision," but where group size was high—thirty-two rats—individuals had trouble learning that another rat needed to be holding down a lever to get water. Calhoun hypothesized that was because almost every time a rat in these groups went to the STAW device, there was another rat there, so they didn't need to *learn* anything to get the water.

Things were even more complicated in the disoperation groups. There, when group size was low—two or four rats—rats had trouble learning that the presence of another rat was bad news for getting water. Here, Calhoun argued that because low group size resulted in another rat rarely being near the STAW, there was little opportunity to pair another rat with an adverse outcome. As group size increased, what the rats in the disoperative groups needed to learn was to time their activities so that they were likely to be around the STAW device when others were not.

With these intriguing results on learning in hand, Calhoun had hopes that, in time, he would get a deeper sense for how group size and learning affected population growth. But largely because of the actions of one troublemaking rat in the disoperation treatment, he didn't get the chance to do that.

In one of the disoperation pens with sixteen rats, some rats were having trouble learning to time their visits to the STAW device so as to be the lone rat there: "An innovation by a single rat [from that group]," Calhoun noted, "opened our eyes to the complexities of social interaction." The twenty-four pens in the STAW experiment sat next to one another but were separated by electric fences, which prevented movement from one pen to another. The innovator rat from that disoperation group of sixteen learned to climb on top of the STAW device, launch himself over the fence, and land in a pen that housed sixteen rats in one of the cooperation groups. Rats in that group had learned that the presence of another rat—any rat—at the STAW device meant a chance for water. "To the [cooperator] rat who joined the [disoperator] rat [at the water levers]," Calhoun wrote, "the [disoperator] rat's behavior was correct. But to the [disoperator] rat, his [cooperator] partner on the STAW was behaving incorrectly."

The innovating intruder would have none of it. "To correct this behavior," Calhoun wrote, "the [disoperator] rat backed out, grasped the offending [cooperator] rat by the rear with its teeth and dragged him out. . . . At no time did the [cooperator] rats fight back despite the pain and wounds they received." This happened over and over, until eventually half of the rats in the cooperation group died from wounds they had received from the innovator, who rather than adopting a "when in Rome" rule, instead played by the rules it had learned before jumping the fence. At that point Calhoun ended the experiment, in part because of the mass mortality in that one group, and in part because if one rat learned to jump the fence, others would eventually learn to as well.[3]

The STAW work was never published, but Calhoun planned to present his preliminary findings to a joint meeting that the American Association for the Advancement of Science and the American Zoological Society were hosting in Denver. The abstract he submitted for his talk—"The Influence of Group Size upon Cooperative Behavior in the Norway Rat"—laid out the basic findings before that innovator jumped the fence, but, in the end, Calhoun just didn't have the time to pull himself away from his work in Bethesda, and so he didn't attend the meeting.

With the unexpected early termination of the cooperation experiment, Calhoun used the additional time to work on setting up and implementing a new experiment that he had designed to look at social behavior and population growth in three ten-by-fourteen-foot enclosures he had built for this purpose. Each of three enclosures was broken up into four equal-sized neighborhoods by two-foot-tall electrified partitions crossing north–south and east–west, so that rats could not scale the partitions and move from neighborhood to neighborhood that way. Every neighborhood in an enclosure had food, water, and nesting material.

Food dispensers in each neighborhood in the three separate enclosures were constructed such that it took time to feed and so that many rats could feed at once, while water dispensers in those same neighborhoods allowed for a quick drink and relatively little interaction with other rats.[4]

Each neighborhood in an enclosure had a housing structure reminiscent of the mouse apartment complexes Calhoun had built back at JAX. Each housing unit was attached to a wall, and to access a building, rats needed to climb one of two spiral staircases to the roof and then enter through one of four openings that led down—burrow-like—to a series of five nesting boxes. Buildings in two neighborhoods would be equivalent to rat high-rises, standing six feet tall, while the other two would be more modest at three feet tall.

Ramps were placed so that rats could move directly between

Each rat enclosure was fourteen feet long by ten feet wide by nine feet high. Each pen (neighborhood) measured seven feet long by five feet wide; the north–south and east–west partitions that created the neighborhoods were two feet high and electrified so that rats could not climb them to move from one neighborhood to another. Ramps connected neighborhoods in the top right (neighborhood I) and bottom right (neighborhood II); bottom right (II) and bottom left (neighborhood III); and bottom left (III) and top left (neighborhood IV). Buildings in neighborhoods I and II were three feet tall, while those in neighborhoods III and IV were six feet tall. Credit: Bunji Tagawa.

neighborhood I to II, neighborhood II to III, and neighborhood III to IV, but no ramp was placed between neighborhood I to neighborhood IV. The upshot of this was neighborhoods II and III could be accessed by two adjacent neighborhoods, but rats could never travel directly between neighborhood I and IV (or vice versa), and so Calhoun referred to them as the "end pens" and neighborhoods II to III as "middle pens."

In his laboratory notes, research proposals, and other writings at the time, Calhoun does not explicitly say that he is attempting to mimic urban neighborhoods in our own species, but the design of the rat enclosures suggests he might have been. And when he was communicating with Eugene Casey about the work that Casey's generous donation made possible, Calhoun turned to human neighborhoods as an analogy. After outlining some recent NIMH meetings on the environment and mental health, Calhoun explained the difficulties of controlled, on-the-ground experiments in people to test ideas in this area. "To do this on a rigorous scientific basis a minimum effort would include building two communities each of at least 100 houses," he wrote to Casey. "In the one you would maximize the esthetic appearance, the arrangement and facilities of a house that enhance development of harmonious family life. . . . In the other 100 houses you would attempt to make all the arrangements and facilities such as to produce dissatisfaction and unhappiness among the people living in them. . . . Then you would want to pry into the most intimate relationships taking place among the people in each of these communities." But Calhoun assured his benefactor, "with experimental animals we could ethically conduct this type of study," and he'd be doing just such things on the second floor of the barn. "Of course, we realize that rats are not men," Calhoun continued, "but they do have remarkable similarities in both physiology and social relations. . . . [W]e can at least hope to develop ideas that will provide a spring forward for attaining insights into human social relations and the consequent state of mental health."

The design of the enclosures allowed Calhoun the power to study neighborhoods that differed in accessibility and the height of the housing units within them. He could then look at how each of these variables affected rat behavior—including aggression, mating choices, nesting behavior, and maternal care—and how these behaviors affected population dynamics and growth at both the level of the neighborhood and the enclosure.

Calhoun intended to dig deeply into all this. The data from Towson and the streets of Baltimore had shown him that dominant rats could be very aggressive. Neighborhoods I and IV, the end pens, were less accessible (one versus two ramps) than neighborhoods II and III (the middle pens), which, in principle, would make them more defensible, allowing dominant rats to set up fiefdoms there. Would such fiefdoms develop? Why or why not? If they did, what impact would they have on population growth? Calhoun speculated that, unlike our own preferences for penthouse living, because more time and effort were needed to go from the nest boxes to the food, water, and nesting material from the high-rise versus lower-to-the ground housing units, rats might prefer the latter. This might introduce what Calhoun called an "income factor" into the experiment with low-rise units being more valuable. Did it? If so, how would it affect behavioral dynamics and population growth?

The answers to these questions and others would surprise Calhoun and set the stage for the research program he developed over the next three decades.[5]

8
PATHOLOGICAL TOGETHERNESS

The Towson enclosure experiment in the 1940s had given Calhoun the power to study population growth and social behavior in a more controlled setting than the row houses of Baltimore. At NIMH, in the Casey Barn, he would have even greater control—as long as he had the rats to get the work started.

Calhoun devoted an entire room at the barn to breeding rats. He wasn't breeding for any particular traits but rather for quantity, allowing him ready access to the many rats he'd need for his work. But there were more to these rats, as Calhoun knew much about the family history of all the individuals he placed into each of his enclosures, as they were descendants of a new NIMH breeding colony that he had established. That line was initiated from the same Parsons Island population that Calhoun had used for the Towson experiment. In the barn he had custom designed "life space" cages to optimize breeding conditions for the male and two females that lived in each. Each life space had food, water, a wheel for exercise, nesting material, and several dark nest boxes into which the two females could withdraw and give birth and suckle their pups away from the male (and each other). It was the pups born in these life spaces in early February 1958 that seeded each of the enclosures on the second floor of the barn.

Healthy female rats begin breeding at three to four months of age, typically producing litters of four to eight pups. The experiment began when sixteen young male and sixteen young female rats, who had just weaned from their mothers, were placed into each of three enclosures with food dispensers that held hard food pellets lodged behind a metal grating. The study was set to run for three rat generations, so that the adult rats at the end of the experiment would be the great-grandchildren of those placed in the enclosures: all in all, Calhoun expected the experiment would take about a year and a half from start to finish.

Out in the Towson enclosure, Calhoun had dealt with the nocturnal lives of his rats by lighting the enclosure enough so that he could observe them at night. The Casey Barn setup provided a much less exhausting solution. He slowly reversed the rats' light-dark cycle so that at the end of the process our day was their night. Then he turned on low lighting during our day so that he could do observations then. Every one to two months, each rat was captured, and data was collected on its size and weight, the number and location of wounds from aggressive interactions, and, for females, pregnancy status. While the rats were captured, Calhoun collected data on the complexity of nests in each apartment building and the amount of feces at each.

To keep track of all that was happening in the enclosures, and to know who was where and doing what, Calhoun affixed a metal tag to each rat's ear and also applied two different colored dyes to each rat's fur so that the rats could be easily identified when researchers were looking down from the overhead windows. Periodically, Calhoun and his assistants would climb the stairs to the subroof, lie down on the three-by-five-foot glass panels with a tape recorder in hand, and dictate the goings-on below, with a special eye for aggressive interactions, mating, feeding, drinking, and nest-building behavior.

The design and setup for the Casey Barn work was unlike anything researchers of the day were using to study population growth and behavior. It was a fusion of ecological, ethological, and psychological thinking that could have come only from someone with the eclectic, unorthodox background of Calhoun. The feeding and drinking devices were reminiscent of large-scale versions of what psychologists used in their work on learning, the censusing and marking techniques were classic ecology, and the sorts of behaviors Calhoun planned to record could be found in any ethology textbook of the day. Whether a melding of such approaches would yield fruit or produce a hodgepodge of unconnected bits was a gamble that Calhoun was willing to take.

From his time working with street rats in Baltimore and those in the Towson enclosure, Calhoun estimated the optimal population size in each enclosure was about forty-eights rats—twelve per neighborhood—so that the thirty-two rats in each enclosure at the start of the experiment would experience only low to moderate levels of stress from interactions with one another. He decided to let the population of adults in each enclosure increase to eighty, at which point he removed any offspring to keep the adult population constant. That meant that, in principle, the population could increase to two and a half times its original population size of thirty-two.

In a *Scientific American* article he wrote based on the Casey Barn study, Calhoun opened by asking readers to consider the forces that political scientist Thomas Malthus had warned checked human population growth in his 1798 book, *An Essay on the Principle of Population*. "In the celebrated thesis of Thomas Malthus," Calhoun told his readers, "vice and misery impose the ultimate natural limit on the growth of populations. Students of the subject have given most of their attention to misery, that is, to predation, disease and food supply as forces that operate to adjust the size of a population to its environment." By design, famine and disease—which fell under Malthus's "misery"—would not be an issue in the experiment that the readers were going to learn about, as there was plenty of food and water and no (infectious) disease to speak of. "But what of vice?" he continued. "Setting aside the moral burden of this word, what are the effects of the social behavior of a species on population growth—and of population density on social behavior?"

What might happen as the rat population size marched upward

to Calhoun's cap of eighty adults? There, he had some experience, he informed his *Scientific American* audience, outlining his work from the Towson enclosure, in which, for rats, there was "no escape from the behavioral consequences of rising population density . . . stress from social interaction." But the Towson experiment had its limits: he could build fences and nest boxes, but he couldn't manipulate important variables like accessibility between neighborhoods and the sorts of housing available. If there really was no escape from the behavioral consequences of rising population density, the Casey Barn rats might help him to understand why.

One of the many things that Calhoun gathered information on was which neighborhoods the rats lived in. Rats were free to move from neighborhood to neighborhood, and did so often, but where did they take up residence? A good measure of that was where they slept, and so that's what Calhoun used as his key indicator of residence. All else being equal, one would expect about a quarter of the population would reside in each neighborhood. But, by design, all else was not equal, as Calhoun had introduced what he called "biasing factors"—the number of ramps into and out of a neighborhood and the height of the apartments in different neighborhoods—to study their effect on population dynamics.

One thing that Calhoun quickly discovered was that because neighborhoods I and IV had only one ramp each, it was much easier for a dominant male rat to control one of those neighborhoods as his own territory, and because neighborhood I had a low apartment complex, dominant males showed an especially strong preference to set up residence there, often sleeping at the base of one of the ramps, controlling access up the spiral staircase. In some cases, the dominant male was the sole male in a neighborhood; in others, one or a small number of subordinate males lived there as well. Subordinates avoided any interactions with the dominant male and, as a result, often fed in other neighborhoods.

Females preferred to mate with dominant males and live in their neighborhoods, in part because the dominant protected them from constant harassment from other males seeking mating opportunities, the upshot of which was that many females would live in a dominant's neighborhood. If a subordinate male was present, he knew his place: "Phlegmatic animals, they spent most of their time hidden in the burrow with the adult females," Calhoun said of the subordinates. "Their excursions to the floor lasted only as long as it took them to obtain food and water. Although they never attempted to engage in sexual activity with any of the females, they were likely, on those rare occasions when they encountered the dominant male, to make repeated attempts to mount him. Generally, the dominant male tolerated these [sexual] advances." In his more technical writings about this work, Calhoun explained how such words as "phlegmatic" were defined, using a term he dubbed "velocity" or "social velocity," which was a mathematical measure he devised to calculate how often a rat engaged in any sort of social interaction.

Calhoun wanted to know about more than just dominant rats setting up their fiefdoms: he wanted to predict how all the rats in the enclosures would distribute themselves under the conditions he had created for them. To do that, he looked at the effect of each of the biasing factors he had introduced, beginning with the effect of the ramps. Working with Clifford Patlak, a mathematician at the University of Chicago, Calhoun predicted that twice as many rats should be found in neighborhoods II and III, each with its two ramps, than in neighborhoods I and IV, with one ramp each. Next, Calhoun assumed that because of the reduction in energy expenditure, the three-foot-tall housing units in neighborhoods I and II would be twice as valuable to the rats as the six-foot-tall high-rises in neighborhoods III and IV, so that twice as many rats would be attracted to neighborhoods I and II as would be to neighborhoods III and IV. Finally, he added together the ratios for each biasing factor and predicted one-quarter of the rats would be found in each of neighborhoods I and III, one-third in neighborhood II, and one-sixth in neighborhood IV.

Between May and September of 1958, Calhoun made 1,386 observations of generation 1 and 2 rats across all the enclosures. The actual number of rats in each neighborhood was almost exactly those he predicted, but because females preferred to reside with dominant males in neighborhoods I and IV, the sex ratio in those neighborhoods was female biased.[1]

Though Calhoun's model correctly predicted the numbers of rats in each neighborhood in generations 1 and 2, as the population increased to Calhoun's cap of eighty adults in generation 3, something very different started happening to the way rats distributed into neighborhoods. The rats slid into what Calhoun called a "behavioral sink"—a term he returned to repeatedly for the rest of his career—that had profound implications for how they behaved in those neighborhoods and how that affected population dynamics. Calhoun shared what happened, and the implications of it all, with the scientific community at two conferences, including the 1959 meeting of the American Association for the Advancement of Science (AAAS). He also published in a technical paper and a very detailed book chapter (based on the AAAS lecture) in an edited volume, *Roots of Behavior*, which contained essays from some of the foremost animal behaviorists of the day. To reach the public, he turned to his *Scientific American* paper.[2]

By generation 3, when the population had reached Calhoun's maximum of eighty adults, in two enclosures, so many rats were residing in neighborhood II—with two ramps and low-rise housing—that whenever a rat fed from the hopper, it was almost always in the presence of many other rats. In fact, Calhoun observed that rats in neighborhood II would rarely eat *unless* there were other rats present. In the third enclosure, the same thing happened, but in neighborhood III (two ramps, but high-rise housing), rather than neighborhood II, which Calhoun noted "is not surprising in view of the indeterminacy of the system," meaning that quirky, chance events matter to some degree.

Calhoun hypothesized that the rats had come to associate feeding with the presence of others "to assure a conditioned social contact," and that this had completely changed the act of feeding: "It reflects a redefinition," he wrote, "of the act of eating as requiring social contact." This was Calhoun's "behavioral sink" that, at the most general level, was an "attraction to one locality to assure a conditioned social contact." That attraction could lead to a "pathological togetherness" in which animals needed to be near others, even if the consequences of such togetherness—eating at crowded feeders when more food could be obtained elsewhere—were negative. It was as if their need for togetherness at any cost sent animals down the drain of a behavioral sink. In his *Scientific American* experiment, it was rats who were sucked down the behavioral sink at food hoppers where, by design, they had to nibble slowly at hard rat-chow pellets placed behind the hopper's grating. As a rat fed, it was almost always surrounded by others who also had to nibble slowly, and so it became conditioned to feed near others (as did those who surrounded it at the feeder). But, in principle, a behavioral sink might emerge in any species at any locality with respect to any behavior.

The questions then became why a behavioral sink would form in the first place and what would happen when it did. Calhoun argued that adult rats fell into the behavioral sink easily because they had learned to associate feeding with the presence of many others when they were one of many pups, all lined up, suckling from their mother. He had also seen this in postweaning juveniles, where "several young rats feed simultaneously . . . they crowd their mouths together as if attempting to gnaw at the same piece of food—this despite the fact that most of the extensive feeding surface remains bare of any rats eating," all of which is to say that Calhoun thought rat natural history predisposed them to slide into a behavioral sink.

The behavioral sink set in motion several vicious cycles. Calhoun argued that because rats were often moving about from one neighborhood to the next, they soon learned that they should feed in neighborhood II (or in one case, neighborhood III). When they did, they'd almost always be doing so in the presence of lots of other rats and so would sate their desire for social contact as they fed. Soon Calhoun was observing sixty of the eighty adults in generation 3 in the enclosure feeding in neighborhood II. At first, all sixty rats were just feeding in neighborhood II, but in time, half the rats took up residence there as well, skewing the distribution away from what Calhoun had predicted, and seen in generations 1 and 2, and producing what he called "pathological togetherness." *Life* magazine learned of that juicy phrase when Calhoun used it in his 1959 AAAS lecture on the behavioral sink, and the magazine immediately used it in their January 25, 1960, "News from the Animal Kingdom" column: "His laboratory rats huddle and overbreed in one pen," the magazine wrote of Calhoun, "even when they have access to neighboring pens with plenty of food. [He] calls this 'pathological togetherness.'"

As more rats settled into neighborhood II (or III), and virtually all of them fed there, the rats' social system broke down. In the crowded neighborhood, "an estrous female would be pursued relentlessly by a pack of males," Calhoun wrote, "unable to escape from their soon unwanted attentions. Even when she retired to a burrow, some males would follow her." Males could be quite aggressive, and females in the crowded neighborhood suffered many more injuries than females who lived with a dominant male in his neighborhood. Females could have left a crowded neighborhood and taken up residence in the neighborhood of a dominant male, who would have been quite happy to have additional females in his harem. But they didn't. It was as if they were infected by the pathological togetherness that the behavioral sink had produced.

Females in crowded neighborhoods were also poor mothers. Building a nest for young is a complicated affair: it involved gathering nesting material—shredded paper from the floor on the enclosure—bringing it back to one of the five nest boxes in their housing unit, shaping it into a cuplike structure with a depression at the center, and sometimes constructing a hood over the nest. Females in crowded neighborhoods failed miserably at this task: "These females simply piled the strips of paper in a heap," Calhoun wrote, "sometimes trampling them into a pad that showed little sign of cup formation." Worse yet, if they encountered another rat as they were gathering up nesting material, these females would simply drop the shredded paper and start interacting with the other rat. Some females in the crowded neighborhoods stopped building nests altogether and just gave birth in the sawdust that was at the bottom of the nest boxes, something that never happened in a dominant male's far less crowded neighborhood.

Mothers in crowded neighborhoods nursed their pups less often than females in less crowded neighborhoods. These females also showed little of the defensive behavior that mothers typically display. In the less crowded neighborhoods, "if any situation arose that a mother considered a danger to her pups," Calhoun wrote, "she would pick up the young, one at a time and take them somewhere safer and nothing will distract her from this task until the entire litter has been moved." If females in crowded neighborhoods picked up their pups to take them to safety—and many times they did not—they often dropped them on the way and left them on the floor. Such pups rarely, if ever, survived. In a dominant male's neighborhood, where mothers built good nests and nursed and defended their pups, 50 percent of the pups survived: Calhoun called these neighborhoods "brood pens." In the crowded neighborhoods, where pathological togetherness reigned, pup mortality reached as high as a devasting 96 percent, the result of a combination of poor nesting and mothering skills, as well physiological deformities of the mother's uterus. Calhoun sent the bodies of eleven females that died in the crowded neighborhoods to a colleague, Katherine C. Snell at the National Cancer Institute. Snell's necropsy report listed severe uterine problems, including endometritis, as well as inflammation of the kidneys, some of which may have been the result of very high levels of vitamin A in the rat chow that all the rats were eating. Calhoun contacted a colleague who was an expert on vitamin A and behavior; this colleague suggested that social stress may reduce the ability of the liver to process vitamin A and that, in part, might explain some of the uterine and liver abnormalities of females in the most crowded pens.

Males were no less susceptible to the effects of the pathological togetherness that the behavioral sink produced. The relative stable dominance hierarchy typically seen in male rats was now constantly in flux. The "free for alls," as Calhoun described them, in crowded neighborhoods, often dislodged those at the top of the hierarchy and replaced them with others. This instability had consequences: it was in the crowded neighborhoods where aggression toward females and pups was highest. In relatively uncrowded neighborhoods with a clear ranking dominant male and few other males, females and pups were usually safe, although even dominant males sometimes "exhibited occasional signs of pathology," Calhoun told readers of his *Scientific American* article, "going berserk, attacking females, juveniles and the less active males, and showing a particular predilection—which rats do not normally display—for biting other animals on the tail."

In addition to dominant males, several other types of males, who far outnumbered the dominants, were seen in all four neighborhoods. "Below the dominant males both on the status scale and in their level of activity," Calhoun wrote, "were . . . a group perhaps [best] described as pansexual." These males, who had lower general activity levels (measured by Calhoun's social-velocity index), attempted to mate with adult females, adult males, and juveniles. Dominant males tended not to resist the sexual approaches of these males, and they rarely attacked pansexual individuals, who put up no resistance when the occasional attack did occur. Calhoun came to think that pansexuality "represents a form of creativity," allowing males to cope with pathological togetherness, in that it was a novel way of "avoiding sanctions from dominant males."

Two other types of males also emerged because of the behavioral sink and its consequences. One type, which Calhoun called somnambulists (sleepwalkers, from the Latin "*somnus*," sleep, and "*ambulāre*," to walk), "were completely passive," he wrote. "They ignored all the other rats of both sexes, and all the other rats ignored them." Somnambulist males showed no interest in mating with females, were largely ignored by dominant males, and no other rats initiated any prosocial behavior, like play, with them. Calhoun discovered these males were the healthiest rats in the enclosures: they were the heaviest, as they spent most of their time feeding, and they rarely had scar marks. The price they paid, Calhoun wrote, was that "their social disorientation was nearly complete," meaning they never attempted to mate or engage in any social action, such as grooming others.

And then there were the prober males, who always lived in crowded neighborhoods II and III. Prober males were the most active rats in any of the enclosures: they were often attacked by dominants, and though they rarely fought back, the aggressive behavior of dominants did not suppress their activity. Calhoun described probers as "hypersexual," refusing to engage in the normal mating ritual, which involves a male pursuing a female, waiting as she goes into her burrow in response, and occasionally sticking his head into the burrow, as well as performing a courtship dance. If the female emerges from the burrow after that, the two mate. But probers would have no part of that: they simply followed a female into her burrow and tried to mate. What's more, because of the high pup mortality in neighborhoods II and III, probers would often encounter dead pups when they entered. Many probers turned cannibalistic at that discovery, which is rare but not unheard of in wild rats.

As Calhoun had planned from the start, after three generations, the experiment came to an end. The population in all the enclosures was still at the maximum size Calhoun allowed—eighty adults—but he was certain that it was just a matter of time, perhaps just a handful more generations, before each population would have crashed and gone extinct. It wasn't just the 96 percent mortality in the most crowded neighborhoods, where most of the females lived, that made him think so. At the end of the experiment, in two of the enclosures, Calhoun removed all the rats except for four six-month-old males and four six-month-old females—rats Calhoun described as "in the prime of life." He then tracked those eight rats to see how they fared: "In spite of the fact that they no longer lived in overpopulated environments," he noted, "they produced fewer litters in the next six months than would normally have been expected. Nor did any of the offspring that were born survive to maturity." As Calhoun saw it, at least at this point in time, there was no escape from spiraling down the behavioral sink: it created pathological togetherness, and pathological togetherness spelled doom for the population.[3]

The press picked up on all this: the *Washington Post* published a long article titled "Gruesome Effects Laid to Overcrowding," and the *Washington Daily News* covered Calhoun's work in a pair of articles—"Overcrowding Caused a Social Breakdown" and "In the City Everyone is a Stranger," which included photos of both Calhoun and his rats. The opening sentences of "Overcrowding Caused a Social Breakdown" read, "The world's population has been growing so fast that social scientists have been studying overcrowded rats for clues to the future behavior of mankind. The man who probably knows more than anyone else about the reactions of pent-up rats is Doctor John Calhoun of the National Institute of Mental Health." The results of Calhoun's work on overcrowding in rats was described as no less than "devasting."

Calhoun told the *Daily News*, "I have two ways of looking at the study of animal behavior. First as animals and then . . . what are the problems at the human level . . . [problems that we] might not even recognize in humans, but once [we do, we] will find it operating in humans." As an example, he turned to the somnambulist males in the barn enclosures. Calhoun told the *Daily News*, in a quote that might very well be cut today by newspapers with a discerning science editor concerned about overextending the implications of nonhuman models, "Perhaps if population growth continues to grow unchecked in humans, we might one day see the human equivalent . . . a sort of withdrawal—in which [people] would behave as if they were not aware of each other. Even beyond that there might be hereditary changes which would produce individuals who would be relatively unaware of their associates," akin to full-time somnambulists. For its part, the *Washington Post*'s piece, "Gruesome Effects Laid to Overcrowding," ends on a rather somber note, reminding readers why they should care about rats at NIMH: "The world's population is increasing at an alarming rate . . . [the hope] is that man will be able to find—as the rats could not find—the political, economic and social formulas to curb it nonviolently, non-horribly."

It wasn't just the press that was paying attention to Calhoun's findings. Anders Richter, science book editor of the University of Chicago Press, wrote Calhoun that he had "read with keen interest and pleasure your article . . . in the February issue of *Scientific American*," and inquired as to whether Calhoun might be interested in writing a book on "natural check(s) built into the dynamics of population(s)." Calhoun was flattered but told Richter he had recently published his lengthy *Roots of Behavior* chapter, in which he provided a more detailed, less-jargony version of his work on the behavioral sink and pathological togetherness, and so he was not inclined to author an even longer exposition just yet. Professors and graduate students, too, were fascinated by Calhoun's findings: at the same time that Richter was reaching out to Calhoun, students at places such as Rutgers University's Institute of Animal Behavior were working through Calhoun's chapter in *Roots of Behavior*.

Not all the attention the behavioral sink and pathological togetherness received was positive. In a 1963 article, "Anarchy in Rat Town," after a positive review of the *Scientific American* paper on pathological togetherness in the rats of the Casey Barn, the author, Peter Broadhurst, made the case that there was no evidence of behavioral sinks in prairie dog "towns" in the great plains of North America, although the social systems of prairie dogs and rats have some striking similarities, particularly with respect to male dominance hierarchies.

Based on his interview with John King, lead researcher on the prairie dogs, Broadhurst told his readers that when population size increases, prairie dogs just move to "new unpopulated areas on the edge of the prairie dog town—to suburbs, in fact." Broadhurst then addressed the question of whether there was evidence of behavioral sinks in nature. What's more, he asked whether there was even an argument to be made for why they should ever be found in nature, answering with a definitive no to both questions. Perhaps the first question was unfair—it takes time for ideas to disseminate through the literature and be tested in other systems—but the second was not: this was a new idea, entirely based on Calhoun's work, and readers would want an opinion on its generality. But Broadhurst appears to have read only Calhoun's *Scientific American* article, not his more detailed chapter in *Roots of Behavior*, where Calhoun addressed that second question head-on. "Certainly, many species have encountered situations leading to behavioral sinks," Calhoun wrote there. Imagine, he proposed in that chapter, a species where small groups of a dozen or so drink at numerous water sources, but slowly, over evolutionary time, their environment becomes drier and drier: "as sources of water became sparser, members of adjoining families or colonies were more likely to arrive" at water holes, setting up conditions for behavioral sink as individuals began to pair drinking with the presence of others. Initially, Calhoun argued, under those circumstances, many pregnant females and pups might die because of the direct and indirect consequences of the large, abnormally high aggregations that come when a population slides down the behavioral sink. What would happen next would depend on whether the behavioral sink worked so quickly that the population crashed to extinction—as Calhoun thought would happen with the rats in the barn enclosures, where the behavioral sink was at its most powerful—or whether "natural selection will favor survival of genotypes capable of tolerating continued association with many other individuals." If the latter, this might lead to "a sequence of circumstances and events [that] forms a plausible path leading to the evolution of herd-type species."

Still, there were questions. Calhoun had seen no evidence that rats were sucked into a behavioral sink in the larger Towson enclosure.

The food was in one large bin, which should have created conditions for behavioral sink. Why didn't he find one? He knew from his field work back on Mount Desert Island, when he was at JAX, that rodents quickly moved into new areas when they opened and were always looking for more living space. Perhaps, given enough space—like in the Towson enclosure, where rats had 10,000 square feet rather the 140 square feet in the Casey Barn—rats might somehow avoid the effects of a behavioral sink that were seen in the barn enclosures. He just didn't know.[4]

Even as he ran the enclosure experiment in the barn, Calhoun knew his days there were numbered. Casey had leased the barn to NIMH for six years, which would have meant that Calhoun would have to pack up and leave in the summer of 1960. Calhoun was temporarily spared that fate: Casey agreed to extend the lease to the summer of 1961 but not longer. Calhoun, working with Paul MacLean, an expert in the brain and behavior who had joined NIH to head a new section in the Laboratory of Neurophysiology, began to think that what NIH needed was not temporary barn space for behavior and population biology experiments but a large and permanent NIH field station, whose purpose, Calhoun wrote, was "to provide a climate in which knowledge concerning the natural history, anatomy, physiology, behavior and sociology of a broad, phylogenetic spectrum can be immediately made available to experimentalists."

Calhoun had experience with many such field stations, albeit on a smaller scale, including JAX's Hamilton Station, and MacLean spent a fair share of the fall of 1959 in Europe learning more about the field research stations in Bern, Seewiesen, London, Oxford, and Cambridge. After MacLean returned, he and Calhoun began pitching the idea for this facility to NIH administrators: depending on the audience they were addressing, they called it the NIH Behavior Research Field Station; the NIMH Behavior Reserve: A Field Station for the Study of Environment, Brain and Behavior; or simply the NIH Farm. Researchers at the proposed field station would include Calhoun (who listed himself as a "social ecologist"), MacLean, an ethologist, a communication ecologist, a comparative social psychologist, a neurophysiologist (all yet to be determined), and, on occasion, a visiting scientist. The station would house multiple buildings (with almost half a million square feet of space), as well as a few thousand acres of land for field studies in ecology. As far as his own work, Calhoun was aiming for about four thousand square feet of lab space, divided into ten rooms, with "catwalks, or some other type of structural devices, which [would] permit the investigator to observe the activities of the subjects." Here, he would continue his work on behavior and population growth, as well as perhaps restart the cooperation-disoperation study.

Administrators at various levels of NIH expressed a serious interest in supporting such a field station as early as 1959, and on May 6, 1960, purchased a 513-acre farm tract—smaller than what Calhoun and MacLean had hoped for—in rural central Montgomery County, near Poolesville, Maryland, about twenty-three miles northwest of the main NIH campus in Bethesda. Soon construction was underway, but Calhoun knew from his experience at the Casey Barn that the mills of the NIH financial gods grind slowly, and it would take years before this station would be ready for day-to-day use. The question became what to do in the interim, between life in the barn and (hopefully) a new beginning at the NIH-field-station-to-be. He decided to use that time "to get away [and] consolidate [his] thinking" on behavior and population growth, ideally at a think tank where he'd be free from other responsibilities and have the chance to do that consolidation surrounded by a brilliant group of scientists who were also there to have time to think. The timing was good. As the barn enclosure experiments ended, Calhoun had been at NIMH for seven years and was eligible for a sabbatical, and what's more, his older daughter, Cat, had just finished grade school, so that a year away would not cause the sort of disruption that it might have otherwise.

Ideally, Calhoun was looking to do his sabbatical at a place that would embed him with a cadre of behavioral scientists, including social scientists who might help develop his thoughts on how best to use animal studies to deal with human population growth. He had left readers of his *Scientific American* article with this final thought: "It is obvious that the behavioral repertory with which the Norway rat has emerged from trials of evolution and domestication must break down under the social pressures generated by population density. In time, refinement of experimental procedures and of the interpretations of these studies may advance our understanding to the point where they can contribute to the making of value judgments about analogous problems confronting the human species." He hoped his sabbatical would provide him time to figure out exactly how. "I expect to pursue even more profitably my studies in experimental sociology," he wrote when applying for his NIH sabbatical. "It is my firm conviction that such studies may produce concepts difficult to obtain with precision from human material. . . . We can develop concepts quickly enough to be of value in ameliorating the forthcoming 'population explosion.'"[5]

The Center for Advanced Study in the Behavioral Sciences (CASBS) in Palo Alto, California, checked every box on Calhoun's ideal think-tank list. Opened in 1954, and funded by the Ford Foundation, CASBS sat on a hill overlooking Stanford University. The area wouldn't be dubbed Silicon Valley until the early 1970s, but it was already home to NASA's Ames Research Center (with its focus on aerodynamics), Hewlett-Packard, Shockley Semiconductor (the first company to make transistors from silicon), and Fairchild Semiconductor (which produced transistors for NASA's Apollo program). There was a buzz in the air: soon Intel and Xerox's Palo Alto Research Center would be located nearby.

CASBS's mission was, as per its charter, "to increase knowledge of factors which influence or determine human conduct and extend such knowledge for the maximum benefit of individuals and society." That broad mission statement was interpreted as a call to bring in not just social scientists but also researchers in animal behavior, who were working on systems that may have application to humans.

Hank Brosin, who Calhoun knew from his days at Walter Reed Hospital, headed the fellowship committee at CASBS, and with Brosin's support and strong letters of reference from John Paul Scott (JAX) and David Rioch (Walter Reed Hospital), Calhoun was offered a year-long fellowship starting in July 1962. The class of fifty-one fellows that Calhoun joined hailed from the United States, Japan, Israel, England, and Austria. Calhoun, whose area of expertise was listed as psychology, joined nine other psychologists, an animal behaviorist, ten anthropologists (many of whom studied nonhuman primates), three political scientists, six sociologists, two psychiatrists, four historians, three economists, two linguists, four philosophers, three mathematicians, a lawyer, and two experts in comparative literature.

In March 1962, four months before he headed to CASBS, Calhoun got a small taste of a think-tank-like environment in California when he joined Leonard Duhl and some of the other Space Cadets at a symposium on the "Environment of the Metropolis" at the Biltmore Hotel in Los Angeles. Rather than present a lecture, Calhoun acted as the moderator and gave the opening remarks for a session on "Urban Ecology." As Calhoun launched into his opening remarks, audience members might have justifiably checked their program to make certain they had not mistakenly stepped into a session on "Philosophical Ecology." "Ecology represents a point of view about life as much as it does a body of concepts," Calhoun began. "An ecologist is foremost a poet and an artist." Speakers at the session, Calhoun informed the crowd, will "paint you a picture of some vibrant essence of reality." He continued, "This morning we, as primitive artists, ask your forbearance. Try to see through our crude pictures to the beauty beyond them." Before introducing Yale's Edward Deevey, who spoke on "General and Urban Ecology," and then landscape architect and fellow Space Cadet Ian McHarg, who lectured on "No Brains: Or, What Man has Done to the Environment," Calhoun only briefly mentioned his rats and his ideas on social velocity. "I will leave to you," Calhoun said to the audience, "any possible interpretation to how these might in any way fit the human species."

Shortly after John returned from Los Angeles, the Calhoun family sold their home—something they had intended to do even before the sabbatical opportunity arose—and, a few months later, headed to Palo Alto. Very much in line with CASBS's mission to use a wide array of disciplines to understand those factors "which influence or determine human conduct," by the time he returned a year later, John Calhoun would no longer be leaving it to others to interpret how his work on behavior and population growth "fit the human species."[6]

9

ODDBALL AND ON-THE-BALL THINKERS

Sitting atop a high knoll overlooking the foothills of the Santa Cruz Mountains around Stanford University, the Center for Advanced Study in the Behavioral Sciences was designed to both foster collaboration and to allow visiting scientists to squirrel themselves away for periods of intense focus, on whatever their area of interest.

When Calhoun arrived, the hub of CASBS was a four-winged building with ranch-style offices that looked out on the hills. No phones were permitted in the offices, and mail had to be picked up at the administrative offices to minimize interruptions. "Just solitude if one wished it," Calhoun wrote of his sabbatical home, "but open doors along the walkways often beckoned one to stop in for a chat with one of the other fellows." Tables under trees surrounding the office building, a centrally placed coffee pot always brimming over, and a communal dining hall all encouraged informal chats and conversations about nascent collaborations and more. Sherwood L. Washburn and Irv DeVore, who were there with other primatologists working on their book, *Primate Behavior: Field Studies of Monkeys and Apes*, would often gather at one of those communal hubs to discuss their budding volume.

Calhoun loved everything about being surrounded by "odd-ball, on-the-ball, and otherwise" thinkers. His year at the center was arguably the most productive year of his career. One day over coffee,

Calhoun struck up a conversation with Harvard's Frederick Mosteller, one of the most prolific and important statisticians of the twentieth century, who authored hundreds of articles on everything from statistical epidemiology and political polling to a very early statistical analysis of baseball. Calhoun mentioned that he and a colleague at NIMH had been discussing building a model to predict the distribution of rats across any number of neighborhoods that differed in terms of the availability of food, the type of housing, and so on. By this time, Calhoun was starting to refer to neighborhoods as cells within a universe, and he told Mosteller than he and his colleague could not solve the math for spaces larger than the four-celled universes that his rats had lived in at the Casey Barn.

Mosteller loved a challenge and had a knack for collaboration, so he and Calhoun replaced their coffee mugs with pencil and paper. In what could have been turned into a motto for CASBS, Calhoun recalls that "three hours and many equations later, the problem was solved." The math was now in place to make predictions about the distribution of animals in a universe with any number of cells. In time, when Calhoun began working with sixteen-celled universes, this equation would become invaluable.[1]

Calhoun's visit to CASBS just barely overlapped with that of John Tukey, a professor of mathematics at Princeton, who also had a long-term association with the Bell Labs. As Tukey was packing to head back home, Calhoun told him some ideas he had on neurobiology and behavior, and like Mosteller, Tukey was interested enough to stop, sit down, and write "page after page of complicated mathematics," which helped Calhoun organize some of his thoughts. Calhoun also befriended CASBS fellows Yehezkel Dror, a political scientist from Hebrew University in Jerusalem, who, a few short years later, joined the RAND Corporation and became famous for his work on "Crazy States," in which fanaticism quashes rationality, and Carl Rogers, the founder of humanistic psychology. For Rogers, a former president of the American Psychological Association, who had recently published one of his best-known books, *On Becoming a Person*, Calhoun's idea of a behavioral sink rang true. Rogers, like Calhoun, worried about the fate of humanity, particularly as more and more people lived in cities. Writing of Calhoun's rats a decade later in "Some Social Issues Which Concern Me," Rogers warned: "The resemblance to human behavior is frightening. In humans we see poor family relationships, the lack of caring, the complete alienation, the magnetic attraction of overcrowding, the lack of involvement which is so great that it permits people to watch a long drawn-out murder without so much as calling the police—perhaps all city dwellers are inhabitants of a behavioral sink, cannibalism and all."

One reason Calhoun looked forward to the freedom that CASBS offered was that it gave him the uninterrupted chunks of time to write that he so craved. Always in search of funding for his plethora of projects, he wrote and submitted two "informal discussions"—white paper reports—to NIH's Department of Health, Education, and Welfare, setting the stage for grant proposals he hoped to submit down the road. One white paper summarized his ideas on optimal group size, and the other—"Induced Mass Movements of Small Mammals: A Suggested Program of Study"—laid out plans for constructing social-physics-based mathematical models based on his Walter Reed work on mass panic.

Calhoun's most ambitious writing project at the center was finishing "The Social Use of Space," a 185-page book chapter that he had been working on for years. Looking back on his year at the center, Calhoun reminisced that it "switched [him] more strongly from a verbal, analytical, sequential and mathematical way of thinking to a visual, gestalt, and poetic modus operandi." Perhaps, but you certainly would not know that from reading "The Social Use of Space." Though Calhoun needed Mosteller's help to solve for the distribution of rats in any sized universe, he had sharpened his math skills considerably by the time he was at CASBS, and the 127 equations in "The Social Use of Space"—equations that he cedes "represent only crude approximations to reality"—made that very clear to readers.

The overarching goal of "The Social Use of Space" was to look at population dynamics through the lens of a mathematical model Calhoun built of optimal group size in mammals: the group size that he defined as "above or below which the altered . . . type of interactions are either stressful or fail to elicit optimum physiological states." Turning to the social physics that he had first learned of in his friend John Q. Stewart's 1948 *Scientific American* article, Calhoun treated each member of a group as a particle moving through space, affecting, and being affected by, other particles in its community. What Calhoun's equations suggest is that the optimal group size in mammals should often, though not always, be about twelve, as in his rats.

What makes "The Social Use of Space" so different from Calhoun's other writings on optimal group size is not just the heavy hand of mathematics but the extension of the discussion of optimal group size to our own species. At CASBS, and to a lesser extent before he arrived, Calhoun did a deep dive into the literature in both cultural anthropology and biological anthropology. Those readings, which he briefly reviews in "The Social Use of Space," led him to believe that modern man had inherited from his primate ancestors a predilection for living in groups of between ten to twenty adults. This optimal group size, he thought, became fixed about half a million years back, long before *Homo sapiens* began to roam the planet, and had not changed since. We still, Calhoun thought, preferred life in small groups of about a dozen to two dozen. That meant he had some explaining to do: if he was correct, how is that we now live in tribes, villages, towns, and cities of far greater size than a few dozen?

The answer was culture, or more specifically, cultural evolution. "Development of a larger social group is made possible," Calhoun wrote, "by a culture in which a normative orientation prescribes values, and sanctions roles of behavior such that the total effect of participation in a larger group so buffers the individual that at any particular time the individual functions socially as if he were a member of a group of 12 individuals." That is, we have developed cultural norms that make us feel and act as if we are in small groups of around a dozen, even when we are not. As actual group size increased, Calhoun argued, our central nervous system evolved to use new cultural rules that made life in the now-larger groups not only possible but created the sense that we are still at our evolutionarily optimal size, which was set long ago. "Human society has developed the form of a many-layered chain link armor," he told readers of "The Social Use of Space." "Each link is composed of not much less than, nor many more than, twelve individuals . . . through time, any one individual shifts his membership back and forth among several joining links. This poetic view embodies the essence of reality."

This process of cultural evolution was not gradual; it occurred in leaps and bounds: at certain points in human evolutionary history—large-scale agricultural, religious, scientific, and electronic revolutions—a new group size was reached that required changes in the central nervous system, leading to a "reorientation of the value system." Then that value system remained stable until the next group size plateau was reached and another value reorientation occurred. Somehow, though, Calhoun made it clear that he didn't know exactly how this evolving, "many-layered chain link armor" protected us from sliding into the behavioral sink.

The problem was that, at some point, population size would become so large that the pernicious effects of the behavioral sink would become inescapable: unless we could come up a radical solution. Perhaps we had already reached the tipping point, Calhoun wondered, and if we hadn't, he was convinced that it was not all that far off in the future, and "without knowledge of evolutionary limitations and universal principles of social physics," any presumptive solution to the population explosion in our species would ultimately prove sterile.

"The Social Use of Space" drew a mixed response from the wider scientific community. In general, reviews of this chapter noted that Calhoun's work on the rats of the Casey Barn was well done and informative. As to his thoughts on human evolution, particularly as it related to his models of cultural evolution and an optimal group size, reviewers were intrigued but quite skeptical. "Portions of Calhoun's interesting, intuitively based attempt at a mathematical formulation of social dynamics appears to have considerable heuristic merit," wrote one reviewer in *Science*. "However, its potential . . . has likely been impaired by overly zealous and stringent anecdotal defense of a conceptual structure that is admittedly founded on secondary assumptions and observations highly susceptible to equivocal interpretation." Another reviewer, this one writing for the journal *Human Biology*, noted that Calhoun's "empirical observations with rats deserve immediate attention," but "the theoretical notions of Calhoun are little more than promissory." As for Calhoun's idea that twelve was the optimal group size in many mammals, including humans, a reviewer for the *Quarterly Review of Biology* was concerned: "As the result of some rather complicated mathematical peregrinations, [Calhoun] eventually arrives at the number 'twelve' as the optimum group size for certain species." The reviewer warned, "This conclusion is derived chiefly from studies of small rodents and, above all, the Norway rat. The extrapolation of these findings to other mammals, including man, will find some resistance."[2]

Calhoun's own worst fears regarding human overpopulation were confirmed when, in one of those dives he made into the literature at CASBS, he came across a 1960 *Science* paper—"Doomsday: Friday, 13 November, A.D. 2026"—that he had apparently missed when it was first published a year earlier. The authors, University of Illinois physicists Heinz von Foerster, Patricia Mora, and Lawrence Amiot, developed a mathematical model that used twenty-four estimates of world population spanning about one hundred human generations, from "the time of Christ" (which they signified as "0") to 1958. Fitting that data to a population growth curve and extrapolating to the future, what they found was that—assuming technological advances allowed for enough food to provide for everyone everywhere and there were no worldwide catastrophes such as nuclear war—on November 13, 2026, the human population would go "to infinity and that the clever population annihilates itself."

Von Foerster and his colleagues understood the sort of simple extrapolations they made were fraught with uncertainty, and they didn't take the model's exact date of annihilation literally. The point was to send out an SOS—one that was based on what we know has happened over the last two thousand, give or take, years of growth. Even if their exact date was off by decades, they argued, there clearly wasn't much time before the population explosion would lead to our demise. Von Foerster's solution was to use legislation to cap reproduction at two children per family and to employ heavy taxation to enforce it. In a reasonably favorable write-up of the warning that the "Doomsday" paper shouted from the hilltops, *Time* magazine could not resist a swipe at the suggested remedy by ending their piece with a single sentence set off from what preceded it: "Dr. von Foerster is the father of three."

Calhoun read and reread his copy of the von Foerster "Doomsday" paper at CASBS and later, underlining this or that sentence. "The logic and beauty of this formulation converted me to its likely validity," he noted. "It laid out with crystal clarity the imminence of the many forebodings by diverse thinkers that the world is entering a unique megacrisis." Like von Foerster, Calhoun didn't take the 2026 date literally, but he was convinced that the study of human populations was now a bona fide crisis discipline, as humanity had decades—at most centuries, not millennia—to figure out a solution to the population explosion problem. From an evolutionarily perspective, for a species like *Homo sapiens*, which has been here on the order of two hundred thousand years and perhaps longer, that's the blink of an eye. "The increasingly fragmented reductionistic pattern of basic research," Calhoun was now convinced, "cannot effectively contribute to the resolution of this megacrisis in so short a time." The solution must come from a radical change driven by a new period of cultural evolution.

Calhoun's ideas on how to foster that radical change were still in their infancy: indeed, he didn't really have any notion of what that radical change might be. Still, he was convinced there were ways that he could move forward on that front. For one, he needed to continue probing population growth in rats and mice and other animals for general patterns that might emerge and possibly be useful in dealing with the human population explosion. In particular, the dynamics of the behavioral sink had only been studied in rats—and there just once, and only for a short time. He could, should, and would, he knew, do much more on that front.

The ideas he developed at CASBS in "The Social Use of Space," Calhoun realized, might also be of value for this new crisis discipline. If he was correct that humans were still physiologically adapted to life in groups of one or two dozen and that cultural evolutionary change produced a milieu in which that "many-layered chain link armor" created the feeling that we were embedded in smaller groups, despite living in much larger tribes, cities, and so on, then researchers needed to get a much better grasp on science and culture in a wide swath of contemporary cultures. We needed more baseline information on what sorts of new cultural norms could, perhaps should, emerge to curtail population growth.

Calhoun thought his studies on population dynamics in animals and his interest in cultural evolution in humans worked synergistically in this new crisis discipline. While he hypothesized that optimal group size in nonhuman mammals tended to be about twelve, he understood that some mammals live in herds of hundreds, sometimes thousands. Natural selection may have favored genetic combinations that somehow spared herd-living species from sliding into the behavioral sink (though some, he thought, likely had and had simply gone extinct). Still, more and more research was demonstrating cultural transmission in animals, and so cultural norms in nonhumans, too, may have been involved in skirting the behavioral sink. "I could not help but ask [myself]," Calhoun wrote in his report of his time at CASBS, "how can experimental research with mammals be most effective in contributing to our understanding of cultural evolution and, in particular, to its culminating aspect, what is popularly called the human population explosion?" No wonder, then, that he described his time at CASBS to his colleagues at NIMH as expanding his "capacity as scientist and citizen."[3]

During his time at CASBS, when Calhoun was not spending time with family, he used whatever spare time he had to write fiction. Even here his mind was never far from the work he was doing at the center. He wrote a poem titled "A: A Mathematical Construct," which was about what he called "conceptual space," and he began writing a bizarre, futuristic, evolutionarily inspired novel called *317 P.H.: A Satire on a Future Multiple "Utopia."* The novel was set in the year 2534, and the "P.H." in *317 P.H.* stands for "*Posthomo*." Calhoun's tale opens in 2034 at a conference being held in Musoma, Tanzania, near Lake Tanganyika. At that conference a decision is made to "direct selection toward the formation of a genus . . . with seven species," who would evolve in Australia as well as Madagascar and other parts of Africa. After some debate about whether these seven new species would be in the genus *Homo*, a decision is made that they would compose a new genus—*Posthomo*. By 2227, selection for these new species had led to the demise of *Homo sapiens* and to the full-fledged *Posthomo* species.

Calhoun only describes three of the seven *Posthomo* species, but each of those has undergone their own version of the sort of radical change that Calhoun thought was needed to save humanity from the population explosion. Calhoun thought that changes would be driven by cultural evolution, but in *317 P.H.* he took the liberty of making such changes genetic, without worrying about the moral implications of such eugenic tinkering.

Each of the new *Posthomo* species colonized a new habitat, allowing their populations to grow afresh. *Posthomo geomys* moved underground and possessed "flattened split nostrils . . . bones extending forward forming a shield around the eyes . . . [and] eyelids fused forming a single lid." *Posthomo maraelephantoides*, an aquatic species with males that weighed ten thousand pounds and females tipping the scales at a trimmer four thousand pounds, was the only *Posthomo* species who, through intense selective breeding, had "achieved the capacity of maximizing gratification to the complete exclusion of aggression and frustration." There was also *Posthomo instersterralis*, a species selected to live on other planets, with individuals who lived up to five hundred years and were capable of hibernation (for interstellar voyages), and who possessed an "internal cybernetic detoxification system, which monitors all metabolic pathways."

Calhoun tinkered with *317 P.H.* for the next twenty-five years but never quite finished it: "Although I have filled notebooks with typed sections," he noted in an autobiographical essay many years later, "I never felt comfortable enough with [its] direction to submit it to a publisher."[4]

When Calhoun finished his year-long fellowship at CASBS and returned to NIMH in August of 1963, many things were in limbo. He and his family temporarily moved into a hotel until they could do due diligence searching for a new house. The construction of the new NIH field station was underway but nowhere near complete. Calhoun expected as much, but what he hadn't expected was that within a few weeks of his return, NIH would send him to Nairobi, Kenya, for a month-long trip to inspect various veterinary stations where mammals were housed.

In Kenya, John and Edith, who was accompanying him on the trip, visited the Ngorongoro District, which was home to Olduvai Gorge, which Richard and Mary Leakey had made famous with their discovery of *Australopithecus boisei*, *Homo erectus*, and *Homo habilis* fossils. While at Olduvai, they dined at the Leakeys' home. After Kenya, it was off to Salisbury, Rhodesia (today's Harare, Zimbabwe), and then to Zanzibar for more inspections. On the way back to the United States, the Calhouns stopped in Frankfurt, Germany, where John lectured on his work on population dynamics in rats at a conference held in the Frankfurt Botanical Gardens.

Shortly after returning from Africa, Calhoun received an invitation from a fellow Space Cadet, Ian McHarg, that would give him a chance to discuss his work on population growth and decline in rodents and his growing concern about human population growth. McHarg, founder of the Department of Landscape Architecture and Regional Planning at the University of Pennsylvania, was an ecologist, a Darwinist, and a landscape architect who railed against the human population explosion that led to our ravaging the planet in search of the seemingly endless land and other resources necessary to sustain our growth. McHarg often invited leading thinkers from a wide array of disciplines to speak to the graduate class he taught on "Man and the Environment." Calhoun was one of fifteen such thinkers who visited McHarg's class in 1963: eight of the other fourteen were, or would go on to become, Nobel laureates, and many appeared on McHarg's weekly CBS television show, *The House We Live In*. In Calhoun's lecture, "Space and Social Behavior," he told McHarg's students about his early work on mice at JAX, but he largely focused on the Casey Barn rats and how their population dynamics resulted in a slide down the behavioral sink. No records exist of Calhoun and McHarg's conversations on population growth in humans and nonhumans during the visit, but those conversations must have inspired both McHarg and his students, as McHarg invited Calhoun back as a guest lecturer many times.

Between CASBS, Africa, and the University of Pennsylvania, Calhoun had been on the road for the better part of fifteen months. He had devoted some of his time at CASBS to planning the experiments he would begin when the field station was ready for him and all the plane time—particularly to and from Africa—to mapping out future experiments. Now back in Bethesda, Maryland, where he purchased a new split-level home in the Wildwood Manor subdivision, just three miles from the main campus of NIH and about twenty miles from the NIH field station, he could devote his full-blown attention to how best to continue his work on social behavior and population growth and decline.

Calhoun had reason to be optimistic and upbeat. Not only were his papers on population dynamics and social behavior being cited in the primary scientific literature and written about in newspapers and magazines, but Julian Huxley (Thomas Henry Huxley's grandson and a central player in the evolutionary synthesis), who Calhoun had met in Kenya, spoke of Calhoun's work in his lectures, and Garret Hardin, who would soon be famous for developing the tragedy of commons model, wrote about Calhoun's studies in one of his books. Administrators at NIH were taking note of all this: a progress report by his superiors noted that Calhoun's "studies are widely referred to in college courses in ecology, psychology, and sociology."[5]

Calhoun made several strategic decisions about what was to come next. For one thing, he'd continue to work with rodents. Though he tinkered with the idea of adding chickens to the mix (studying their behavior with an eye toward understanding population dynamics), for logistic reasons, he decided that was just not feasible. To probe deeper into the population dynamics and the resultant behavioral sink that he had found in rats in the Casey Barn, Calhoun was contemplating switching back from rats to mice. But he was not about to completely part ways with the rats that had taught him so much: once he was set up at the field station, he hoped one day to eventually rekindle the STAW cooperation study—this time designed so that no rat would be jumping over any fences and causing havoc.

After interminable delays of one sort or another, by late 1965, the NIH Behavior Research Field Station was ready for occupancy. And so, early each morning, John Calhoun jumped into his Opel car and drove twenty miles northwest, much of it on dirt roads and through forest, to his new home away from home.

10

THE RANTINGS OF A MAD EGGHEAD LOCKED IN HIS IVORY TOWER

Calhoun was excited to have a permanent place to set up shop, knowing full well that was never a given: John C. Eberhart, the director of NIMH, was fond of noting that "it has been said that around the NIH space is love, and that the repercussions of awarding or withholding it are as profound." Still, there was much work to be done before data could be collected. A staff needed to be hired and equipment ordered. Starting in 1966, Calhoun was given four thousand square feet of temporary workspace, but a new building with eight thousand square feet of space, scheduled to open in early 1967, was being constructed just for him.[1]

Calhoun resolved that once he was in his new building, he would continue mapping out future work on cooperation, learning, and population growth in rats (the STAW experiment). But these experiments with rats, as important as they were to Calhoun, would not be his mainstay in the NIH field station, as he had come to the decision, once and for all, he would shift the lion's share of his work on population growth and crashes to mice.

Part of the decision to work with mice was logistic. Mice are smaller than rats, which meant Calhoun could house more universes with more cells (neighborhoods) full of mice than rats. Mice reproduce more quickly than rats, which meant more generations per year.

Mice also produce more offspring per litter, so populations could grow more quickly—and perhaps slide into the behavioral sink more quickly. Calhoun had worked extensively with mice at JAX, and so he didn't need to deal with the learning curve that comes with working with a new species.

What really sealed the deal for mice were the interactions Calhoun had with Alexander Kessler, a PhD student at Rockefeller University in New York City. When Kessler began his dissertation work in 1961, he already had an MD from Harvard and had done a residency at the College of Physicians and Surgeons at Columbia University. But Kessler had developed an interest in both genetics and ecology and so joined Rene Dubos's lab at Rockefeller to work on his thesis, "The Interplay between Social Ecology and Physiology, Genetics and Population Dynamics of Mice." Kessler knew of Calhoun's work, likely from his own readings but, if not, certainly from Dubos, who just a few years later cited Calhoun's *Scientific American* article in chapter 3 of his Pulitzer Prize–winning book, *So Human an Animal*.

Kessler wrote Calhoun for advice on designing an experimental setup for his work, and Calhoun sent him sketches he had been developing for his own work with both rats and mice. Their initial exchanges had a real impact on Kessler: in his thesis, he thanked Calhoun in the acknowledgments, cited eleven of Calhoun's papers, as well as his monograph *The Ecology and Sociology of the Norway Rat*, and in his "Materials and Methods" section, Kessler noted he used "a modified version of enclosures used by J. Calhoun for rat and mouse populations."

Once Kessler's experiment was up and running, he invited Calhoun to Rockefeller to show him around. Kessler's focus was different from Calhoun's, as he was primarily interested in whether selection for individuals of larger size changed population dynamics (it didn't) and whether there was evidence for changes in the frequency of genes linked to coat coloration as population size increased (there was). Still, Calhoun was certainly pleased that Kessler was finding signs of a behavioral sink in his work. As with Calhoun's own work

John Calhoun and a sixteen-cell universe. Photo by Calvin Nophlin/National Institute of Mental Health. Photo credit: Calhoun, J. B. 1969. "Design for Mammalian Living." *Architectural Association Quarterly* 1: 24–35.

in rats, in Kessler's mice, as population increased, pregnancy rates declined, neonatal death rates rose, and there were instances of cannibalism, and while male fighting in general decreased, instances of "unusual aggressiveness" increased. The mice, though, differed from Calhoun's rats in an important way: if mice were taken out of a population that appeared to be sliding down the behavioral sink and placed in more pleasant conditions, they tended to revert to normal behaviors. Calhoun's Casey Barn rats hadn't.

What struck Calhoun most during his visit with Kessler was the sheer size of the populations, even in enclosures smaller than what he had suggested to Kessler early on. In one of Kessler's enclosures, the population had reached one thousand, and in the other, eight hundred. The density of mice was mind blowing. Eighty-five mice per square foot in the larger population: "standing room only" is how Calhoun described it. He couldn't stop thinking about those numbers and the growth rate that led to them: that sort of reproductive potential in mice, paired with the many universes he could build with the space and resources he had at the new field station, would allow him to study behavioral sinks and population dynamics in a way that was simply not possible with rats. "In a world beset by three major crises that threaten to do man in, corruption of his habitat, nuclear warfare, and overpopulation," Calhoun noted in one of his annual NIMH progress reports, "it is fitting that all people direct their energies to preventing any of the three from moving beyond the tolerance point." To that end, his laboratory would now focus on "the study of populations not in humans but in mice."[2]

Calhoun knew how *his* mouse universes would be created. He'd vary the initial population size and the size of the universe. "Everything . . . was to be Utopian," he wrote. "Each universe an island box unto itself contained. Each universe a walled enclave of cells. Each cell . . . a 16-unit high rise apartment." As with the rats in the Casey Barn, to facilitate observation, Calhoun would reverse the mice's light-dark cycle and turn on low lighting during our day to do observations then. There would be "an always replete cafeteria, open spaces for ambling and interacting, construction materials for building nests, [and] many choices for route and action." And the "Spectre of Death [was] excluded: no inclement weather, no predators, no epidemic disease, no famine, no emigration to produce loss of members."

Calhoun and one of his assistants, Jerry Wheeler, were no less than "euphoric in planning of [the] experiment" and began constructing two-, four-, eight-, and sixteen-cell universes for what Calhoun called Study 102. The walls of these square universes were 54 inches high, and each cell in a universe contained one apartment building, covered in wire mesh. Each building had four floors, each with four apartments (nest boxes). To enter an apartment, mice had to climb up a tunnel; once inside an apartment, they could look out over the universe through a circular hole cut out of the mesh. A wire-mesh feeding station was attached to the right side of the building and contained pellets of food that the mice could nibble on. The food station was constantly replenished by researchers, who could access it from behind the wall of the universe. At the top of a building sat four 85-ounce water bottles. Nesting material for an apartment was available in a can sitting on top of a small pole on the cell floor, and, depending on the experiment, also on the floor itself.

Each cell in a universe had 640 square inches of floor space associated with it, which meant that universes with more cells were larger than those with fewer cells. A four-celled universe in which walls ran 51 inches long (for a total of 2,601 square inches) was just a shade larger than 640 square inches for each of the four cells. Four-celled universes were divided into quarters by partitions that were about 3 inches tall: while mice could climb over them, the partitions generally served to demarcate a group territory for the mice living in an apartment building in a cell. The walls of a sixteen-celled universe ran just a shade over 100 inches long, for a total of approximately 10,000 square inches, or about 640 square inches for each of the sixteen cells. Here, partitions were laid in a radial form, as in the spokes of a wheel. Regardless of the size of a universe, the outer walls had steps that Calhoun and his assistants could climb to observe what mice were doing.[3]

Calhoun enjoyed the challenge of shifting a large chunk of his research to population dynamics in mice. Around this time, he was part of a Space Cadet's meeting in which he and twenty Cadets met with Peace Corps volunteers who had recently returned from their tour of duty in far-off places. The ostensible goal was to advise volunteers on "sound mental health planning," but Calhoun couldn't help but think of the Peace Corps volunteers as a population and, hence, subject to analysis like any other. The Peace Corps population was the antithesis of a technical society in which "standardization, order, predictability, hierarchy, minimum channels of communication," ruled supreme. The Corps represented what Calhoun called alter-technology:

"programmed diversity, programmed uncertainty, sufficient unpredictability and uncertainty[,] . . . an everchanging process[, and] . . . permanent adolescence." Given the choice, Calhoun preferred that alter-technology—both personally and in terms of the thrill it injected into the science of studying populations. "I am approaching 50. . . . I now can tolerate all my peers who are now set in their ways, or more set than I am," he wrote, "so I am glad to be a sort of permanent adolescent." Shifting his research program to mice was just the sort of thing that made science exciting.[4]

As Calhoun's building, as well his universes within it, were being constructed, and as he designed the experiments to take place in them, he began a project that would end up taking nearly twenty years to complete. The idea was to construct a very general giant database on published material relating to questions of population dynamics and mental health. That database already ran 450 pages in 1966, and Calhoun was planning on inviting hundreds of people to contribute to a published anthology on the topic. Calhoun hoped that the anthology would be a paper equivalent to the sort of information hub that would become available to tap into by anyone some day in the future. Calhoun also used the construction time on the new building to publish, including penning a review of his work on rodents and population dynamics in a special issue of the *Journal of Social Issues* devoted to "Man's Response to the Physical Environment." When not overseeing construction or writing, he was often on the road lecturing.

In 1965, Calhoun presented a talk titled "A Glance into the Garden" at Mills College in Oakland. In it, he spoke of his work on social velocity (how often individuals interacted with others) and optimal group size in both rodents and humans. Calhoun also used the lecture to suggest what he thought, or at least hoped, would be the next step in the reorientation of our value system: a reorientation system that might help us deal with human population growth. The last four human value reorientations, Calhoun proposed to the Mills audience, centered around the agricultural, religious, scientific, and electronic revolutions, all of which involved "understanding, creativity and communication in their completion," but, he added, "compassion as a universal characteristic has been sadly missing." Calhoun understood compassion as a composite of reason and emotion, a mechanism for understanding "our fellow man . . . and both the limitations and potentialities of our being and our action," and he proposed, without much justification, that the "compassionate revolution" would be the next reorientation of our value system and, if he was correct, would prevent us from sliding into the behavioral sink.

In addition to describing in relatively broad strokes the compassionate revolution, Calhoun suggested something else that might help us avoid self-destruction through overpopulation. To do that, he introduced his listeners to system scientist Sir Geoffrey Vickers's idea of an appreciative system, which Calhoun summarized as "a network of communication so structured as to enhance the likelihood of its rapidly leading to a consensus of opinion among its members regarding goals for human action despite initial divergence of viewpoints." He was certain the rats in the Casey Barn had a thing or two to teach humans about such appreciative systems. The key was Calhoun's belief that optimal group size in both his rats and his mice and in humans was twelve. "My studies on rodents have suggested a means for developing appreciative systems capable of circling the globe and evolving new images of destiny without coercion," he noted. "In essence this involves a network of 'invisible colleges.'" These invisible colleges could be used to discuss and debate, in a critical fashion, important issues such as the human population explosion.

At the global level, Calhoun proposed the creation of twelve invisible colleges, each focusing on a different topic: science, economics, health, and so on. These invisible colleges were to be like think tanks, in that there would be no long physical buildings, no administrators, no set curricula, no courses, no granting of degrees, nor any of the other things associated with a classical physical college. Indeed, the term "invisible college" had been used as far back as the mid-1600s, when members of what would become the Royal Society of London first gathered. Each of Calhoun's twelve invisible colleges would be made up of twelve groups, with twelve individuals per group, each a recognized global leader in their area of study.

In Calhoun's unpublished "unofficial directory for a new decade," which he referred to as "a directory of the invisible colleges," he suggested that members of the twelve invisible colleges might be drawn from "Sociology, Psychology, Architecture, Social Psychology, Psychiatry, Anthropology, Statistics, History, Public Administration, Political Science, Economics, Systems Analysis, Ecology, Agriculture, and General Systems Theory." College and group members could communicate in any way they deemed fit, including in-person meetings. Each group in the twelve invisible colleges would be responsible for seeding a branch of that college, leading to 144 multinational branch colleges: twelve dealing with science, twelve with economics, twelve with health, and so on. These multinational branches would be home to experts, though not necessarily globally recognized leaders, in the subject matter of that college.

Each of the groups in the multinational invisible colleges would seed a national branch. Members of those 1,728 (12^3) national branches of the invisible college would be regional leaders in the subject matter of that branch. Each group in a national branch would seed a metropolitan branch, 20,736 metropolitan branches (12^4) in total. Finally, each group in a metropolitan branch would seed a village or community group, made up of twelve individuals—locals interested in the subject matter—for nearly three million participants at the village/community level. Then, relying on an analogy that Bernard Muller-Thym had made about appreciative systems in commerce, Calhoun suggested that invisible colleges and branches and subbranches would act as nodes in a global brain.

H. G. Wells had popularized a similar notion in his 1938 book, *World Brain*, a collection of his lectures in which he opined on "constructive sociology . . . a subsection of human ecology." Wells introduced the idea of both a world library and a world brain that he thought had the potential to bring about global peace through the free exchange of information and ideas. Just a few years before Calhoun lectured on invisible colleges, Arthur C. Clarke, in his 1962 book *Profiles of the Future*, predicted the emergence of a world brain by the year 2100. And so, while the idea of some sort of global brain per se was not new, nor was the term *invisible college*, Calhoun's ideas on invisible colleges, based on optimal group size in rodents and humans, were.

The Mills College talk (and the book chapter that followed from it) gave Calhoun the chance to introduce his rodent-inspired ideas on appreciative systems and invisible colleges, but he knew his talk and book chapter would only reach a small audience. However, a lecture he gave at arguably the most important annual gathering of scientists in the world—the American Association for the Advancement of Science (AAAS) meeting in Berkeley, California—gave him the chance to reach many more.[5]

At that AAAS meeting, Calhoun was one of five speakers in a session—"Nets: Social, Neuronic, and Others"—that W. Ross Ashby, an early pioneer of cybernetics, had organized. Calhoun opened his talk with a discussion of the Towson enclosure rats and the Casey Barn rats and told the story of rat creativity he had witnessed in the Towson work. There, he told his audience, he had seen one group of rats do something that he thought of as the equivalent of the creation of the wheel in our own species. Normally, rats clear out a burrow one mouthful of dirt and debris at a time. But this creative lot of rodents formed a large ball of debris and *rolled* it out of a burrow: a much more efficient solution.

Calhoun talked of creativity in the context of his idea of social velocity. Switching from the Towson to the Casey Barn rats, he noted that the high-velocity rats—dominant males and females—were more perceptive, in the sense of responding to normal day-to-day conditions. But Calhoun proposed that subordinate, low-velocity rats—who by definition interacted less often than others, and who were not subject to "normal social pressures" and had "little awareness of their social milieu"—might very well use that freed-up time creatively and be better suited to handle a dramatic change from normal day-to-day conditions. "This formulation of creativity must stand for the present as an inference." Calhoun also said, "Although my observations have dealt strictly with rodents, there seems to be no inherent reason why the general formulation is not equally applicable to the human situation."

Next it was time for Calhoun to veer "from this firmer path, paved with data, and enter upon the frail rope bridge . . . to search for possible processes . . . within the context of human relations." Calhoun told his AAAS audience of Vickers's ideas on appreciative systems, and of his own idea that optimal group size in our species is around twelve, citing data on the "Bushmen" of the Kalahari Desert (today, properly referred to as the San peoples). He then introduced the audience to his invisible colleges made up of interconnected appreciative systems with nearly three million participants, a figure he knew would turn heads: "Upon first inspection such involvement of three million people in discussion groups," he told the AAAS audience, ". . . may sound like rantings of a mad egghead locked up in his ivory tower and out of contact with reality."

When Calhoun explained his ideas on appreciative systems, global brains, and invisible colleges in a progress report the year after his AAAS talk, he made it all too clear that what he had in mind would take time, progress in an uneven fashion, and meet with resistance: "Men and institutions will view these developments at any stage as both threats and opportunities, if they note the processes of change and their portents at all," he wrote the higher-ups at NIMH. "There will be struggles, disasters, victories—unevenly distributed and unevenly apprehended by the winners and losers." But Calhoun thought that something that bold was needed to save us: it was from these village and community groups on economics, science, health, and more that he hoped the radical ideas on how to save us from the depredations of uncontrolled population growth, as well as other pressing issues, might emerge.[6]

Calhoun wasn't the only one talking about his ideas on humans and his research on rats: everyone from biologists, psychologists, and anthropologists to city planners were starting to do so with greater frequency. The year after Calhoun gave his talk at the Berkeley AAAS meeting, a committee charged with revising parts of the building code in Philadelphia met. The last item on their agenda dealt with regulations on the minimum number of square feet per person in city dwellings. Some committee members argued that with advances in modern appliances, heating, and electricity, a reduction of the square-foot requirement in the building code was reasonable. Another committee member asked what was known about "the sociological and psychological aspects of varying the amount of living space." The committee was told of some studies from France that suggested reducing living space produced "a deterioration of the relations among the contained members of a dwelling unit," but the committee wasn't convinced. Then someone spoke about of Calhoun's *Scientific American* paper on the Casey Barn rats, and that did the trick: the committee opted to increase the square-foot requirements. A few years later, in an article about population growth and birth control, Dr. Mary Steichen Calderone, the first female director of Planned Parenthood, who had recently cofounded the Sexuality Information and Education Council of the United States, wrote of the denizens of the Casey Barn: "Calhoun in his rat population studies has demonstrated all kinds of abnormal behavior patterns that follow crowding."[7]

The Casey Barn rats were also mentioned in Lewis Mumford's book *The Urban Prospect*. In the book, Mumford, who the *New York Times* described as "philosopher, literary critic, historian, city planner, cultural and political commentator, essayist and perspicacious writer on the subject of architecture," discussed everything from the population explosion to the growth of suburbs—what he called the "slurban explosion"—to the dangers of megalopolis cities. Toward the end of *The Urban Prospect*, Mumford tells his many readers that "no small part of this ugly urban barbarization has been due to sheer physical congestion: a diagnosis now partly confirmed by scientific experiments with rats—for when they are placed in equally congested quarters, they exhibit the same symptoms of stress, alienation, hostility, sexual perversion, parental incompetence, and rabid violence that we now find in Megalopolis."

Anthropologists also turned to Calhoun's work. Calhoun had kept in touch with Edward Hall, ever since Hall had come to lecture back in the Walter Reed Hospital days. Since their first meetings, Hall's ideas on proxemics—the relation between people and physical space—had become widely known, and in 1966, he published a popular book on the topic: *The Hidden Dimension: An Anthropologist Examines Man's Use of Space in Public and Private*. A glowing review in the *Chicago Tribune* called it "one of the few extraordinary books about mankind's future which should be read by every thoughtful person." Fourteen of the sixteen chapters are, as one might expect, focused on humans and space, but after an introductory chapter on "Culture as Communication" and then a general chapter on "Distance Regulation in Animals," the majority of chapter 3 is devoted to Calhoun's work with rats. Hall taught his readers about the Towson enclosure work and the Casey Barn experiment, introducing them to the behavioral sink, hyperactive males, probers, and pansexual rats, and he described the work as "sufficiently startling to warrant a detailed description." Pay attention, Hall told his readers, "The findings of these studies are so varied and so broad in their implications that it is difficult to do justice to them. They should continue to produce new insight for years to come."[8]

In spring 1967, the same year that he lectured on his work on behavioral sinks in the Casey Barn rats to the Department of Behavioral Science at the Harvard School of Public Health, Calhoun moved into his new building at the NIH field station and quickly began an experiment in his (and the mice's) brand new universes. On December 30, 1968, he introduced the scientific world to the first data on a mouse universe at the AAAS meeting in Dallas. Calhoun was one of a handful of keynote speakers at the meeting, and his photo was one of only a dozen that were used to advertise the meeting to thousands a month earlier. The symposium that Calhoun was part of was organized by Aristide Esser, a psychiatrist and proponent of the fledgling field of environmental psychology. Initially, Esser had lined up Konrad Lorenz, the most famous animal behaviorist in the world (who, a few years later, would receive a Nobel Prize for his work), as the keynote speaker. But, when Lorenz had to cancel, in a show of just how much attention Calhoun's work on behavior and population dynamics was garnering, Esser turned to Calhoun to replace Lorenz.

Calhoun's lecture wasn't anything like the typical stodgy keynotes at AAAS annual meetings. He opened with a discussion of his own family history, telling a no doubt mystified audience how 150 Calhouns had come from Scotland in 1760, the result of "general tension accompanying the incipient population explosion which resulted in part from a decline in the authority of clan chiefs to control marriage." He explained that one lineage produced fighters like secessionist John C. Calhoun but that he was a descendant of the other branch, which produced teachers who retreated into the world of the mind. Next, he answered the question that must surely have been on the minds of most of the audience: "Whether this family history has any bearing on my specific scientific concern with the subject of space, I do not know," he said. "However, it exemplifies . . . that there are two kinds of space, one physical and one conceptual. If we are to grapple successfully with the myriad crises and tensions accompanying the developing rapid increase in human numbers, both physical and conceptual space need to be considered."

From there, Calhoun spoke about the ongoing studies of two-, four-, eight-, and sixteen-celled mouse universes that he had begun in July 1968. Each universe was seeded with four males and four females. Calhoun described the construction of the universes and, based on some pilot work, explained that *each cell* (with its sixteen apartments) could support about 250 mice before "standing room only" conditions would put a break on population growth. What he ultimately wanted to know, Calhoun told the crowd of scientists, was what permitted or prevented the mice in that universe from reaching that upper limit (e.g., 1,000, in a four-celled universe). But it was too early to know that, as the universes were young and studies were still in progress.

Calhoun suggested to the audience that with feeders designed so that mice had to nibble at the pellets slowly, a behavioral sink would eventually emerge, as with the Casey Barn rats, though it hadn't quite yet. What he did know, and shared that day, was that starting with eight colonizers, the larger the size of a universe (the more cells within and hence the more floor space), the *more* likely it was that males formed territories and the *longer* it took before the successful rearing of any litters. Calhoun knew he would need to posit a theory for these counterintuitive results—shouldn't *more* space lead to males being *less aggressive* (plenty of space for all) and females producing and caring for young earlier? What he proposed was that in larger universes the initial four male colonizers could roam freely over lots of space and rarely encounter one another. As they roamed, they began to associate the space itself and the inanimate objects they encountered as part of *their* space. "All of these objects with which each mouse gained identity became part of his extended self," Calhoun explained. "Each individual's body by such measure became much larger," because, as Calhoun interpreted it, a mouse's perceived notion of self included the space and inanimate objects within its space as an extension of its physical body. "Two individuals could thus collide even though their physical bodies were some distance apart," Calhoun told the audience, "in so long as one of them occupied or passed through a space that the other had come to identify as part of himself." Once this territoriality was in place, when males did encounter each other, very nasty fights followed.

It was different in smaller universes. There, Calhoun proposed, males were almost always learning about their environment in the presence of another male: "for all practice typical purposes, no individual identified any of its surroundings as being part of itself," and so fights were rare, and when they did occur, they were not especially violent. As for the females, in large universes, where males formed hierarchies and were territorial, "the overflow of stressful situations affecting [them] . . . reduc[ed] conception and interfer[ed] with proper maternal behavior." In smaller universes, this never happened. Physical space, through its direct and indirect effects on male and female behavior, Calhoun told the crowd, affected population growth in ways that revealed new, counterintuitive phenomena.

Calhoun's ideas on territoriality came at this phenomenon from a different perspective from that which was developing in the fields of animal behavior and ecology. For Calhoun, it was the frequency of interaction between individuals that was paramount. But some researchers were beginning to think that it was the distribution of resources in the environment—in particular, whether resources were found in clumps and patches or were more evenly distributed across the environment—that determined whether territoriality was favored by natural selection. When resources like food, mates, and safe spots are patchily distributed, and therefore economically defensible by a putative territory holder, territoriality was favored. Otherwise, it wasn't. That perspective, which is widely accepted today, was not antithetical to Calhoun's in the sense that patchily distributed resources would decrease the frequency of interactions among animals. The difference between these views of territoriality lay not in the frequency of interactions, per se, but in the importance of how resources are distributed.

Calhoun's 1968 talk also saw the return of the creative rats of the Towson enclosure, but this time Calhoun provided a bit more information than he had in the lecture he had given at the AAAS meeting in Berkeley. The group that came up with the rat equivalent of the wheel, it turns out, was one composed strictly of subordinate males who had left other groups as the result of the excessive violence they experienced. These males had all displayed low social-velocity scores, meaning they engaged in relatively few social interactions: but when they did interact, they could come up with creative solutions to problems (in this case, clearing a burrow) that rats in other groups could not devise.

Calhoun went on to say that he had seen other acts of a creative nature in socially withdrawn rats—though he does not say what those acts were—and that the link between low social velocity and creativity fascinated him. This led him to measure the behavioral sequences (grooming, sleeping, eating, drinking, and moving) of high- and low-velocity rats. What he discovered was that high-velocity rats stuck religiously to an ordered sequence and rarely varied from it, a regime that Calhoun argued worked in stable environments. But low-velocity rats exhibited a more random sequence of behaviors, such that one could not predict which behavior would follow which. That randomness, of course, usually led to maladaptive solutions. But, particularly in environments undergoing rapid change, Calhoun proposed, more random sequences might occasionally lead to creativity: low-velocity individuals might, on occasion, stumble on new solutions, some of which might allow for survival in environments where populations were growing at alarming rates, and this could apply to humans. "To the extent that these insights may have applicability on the human level," Calhoun told the AAAS audience, "it necessitates placing a new perspective on the general issue of mental health." That new perspective involved value judgments about what defines a creative solution, as well as tolerance for the many mistakes along the way: Calhoun was convinced his rodents suggested it was worth thinking deeply about all this.

After discussing life in the mouse and rat universes, Calhoun told the crowd of von Foerster's doomsday population model that still obsessed him, as well as of appreciative systems and the idea of the compassionate revolution. Here, Calhoun linked all three together in a way he had not in the Mills College talk. He told his AAAS audience that he assumed humans could sustain a population of 9 billion, and things would be very dangerous at 13.5 billion and perilous at 18 billion (in 1968, the world population sat at around in 3.5 billion).

To keep the population explosion under control and avoid doomsday, he argued, "required an augmented awareness of the necessity for others to maintain value sets differing from one's own . . . [and] realizing one's own functional role requires expenditure of considerable effort in assisting others to fulfill the objectives of their value sets." In other words, altruistic behavior paired with, and facilitating, a wide diversity of ideas. "It is this awareness of, and participation in, the realization of values held by others," Calhoun summarized in his sometimes-cryptic wording, "which characterizes the compassionate perspective."

The compassionate revolution that Calhoun hoped would save us from the population explosion and the revolutions (reorientation of value systems) preceding it were presented on a slide of what Calhoun called a "conceptual homunculus." "The body of man has remained his earlier biological self," he told a group of scientists who were not used to seeing such slides or hearing such language at one of their most important annual gatherings, "but the degree to which he has effectively utilized his cortex has continually increased. . . . This figure indicates the degree of enlargement of conceptual target diameter at the time of each of the revolutions."[9]

The problem, as Calhoun saw it, was that the conceptual space associated with compassionate revolution was so large our brains needed help. His solution was the creation of what he called electronic prostheses. "We become linked," he proposed, "to these to permit further enlargement of our conceptual target diameters beyond the limitations imposed by the cortex." Harkening back to his vision of human society as a "many-layered chain link armor," these electronic prostheses, Calhoun proposed, created a massive conceptual space that would remediate the decrease in physical space per person as our population skyrockets out of control. "To continue enlarging conceptual space requires involving more and more individuals in a common communication network," he said. "Such union will continue

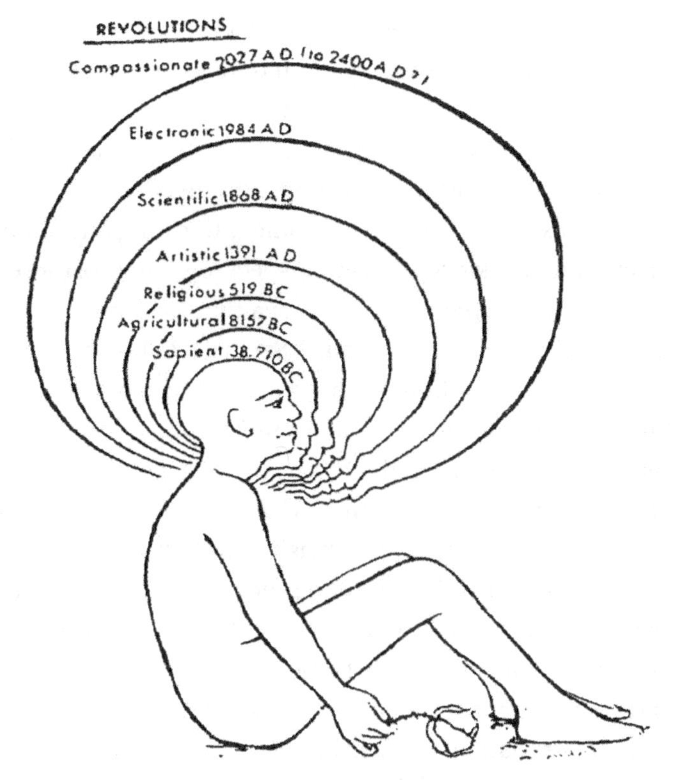

Calhoun's "conceptual homunculus" as it relates to various conceptual revolutions. Credit: John B. Calhoun, "Space and the Strategy of Life," in *Behavior and Environment*, ed. A. H. Esser (Boston: Springer Nature, 1971), 387.

until the entire world population becomes incorporated into a single network." In a brash extrapolation, Calhoun stood before the AAAS crowd that day and predicted that if his calculations were correct, eventually "more than 99% of the conceptual target diameter of the average individual will result from activity within thinking prostheses to which he is linked." Extrapolating from prior conceptual revolutions, he proposed that would happen in the year 165,340 CE.

Calhoun does not explicitly describe what the electronic prostheses were or how they would work, but his AAAS discussion of science fiction writer Leo Szilard's 1949 story "Calling All Stars" provides some hints. Subtitled "Intercepted Radio Message Broadcast from the Planet 'Cybernetica,'" that story revolves around the residents of Cybernetica, who "observed on the earth, flashes which we have identified as uranium explosions" and noted that "uranium is not ordinarily explosive." There are some interesting bits about natural selection in this story, but with respect to brain prostheses, it is the structure of societies on Cybernetica that Calhoun focused on. "Our society consists of 100 minds," Szilard's Cybernetican storyteller says. "Each one is housed in a steel casing containing a thousand billion electrical currents." In his talk, Calhoun calls those one hundred minds "one hundred thinking prostheses." Calhoun's electronic prostheses, then, were akin to the key nodes in a global brain that was made up, in part, by his invisible colleges.

There is no record of the size of the audience at Calhoun's lecture, but given that it was heavily advertised and listed as part of a "Frontiers in Science" forum, it is probable that the crowd was large. We don't have a transcript of the real-time responses from the audience, but since a large swath of those in attendance at AAAS meetings were from the "hard sciences" (physicists, chemists, etc.), and considering that many of the biologists present were not ecologists, population biologists, or evolutionary biologists, it's likely there was many a skeptic in the audience, alongside some who thought that—aside from bits here and there on the behavior of rodents—what they were listening to wasn't science at all. On the other hand, when Murry Chastain, an editor at Harvard University Press, learned of Calhoun's talk, he wrote Calhoun that this "important topic could become a short book that would be appropriate for [Harvard University Press] to publish." Calhoun was flattered, but not ready to take that step, and wrote back: "I make progress on these matters very slowly and do not wish to make any commitment as of now."

All preconceived notions of boundaries had been dropped. Calhoun was moving seamlessly from the population dynamics in rats and mice, which he still sincerely believed were important for their own sake, to the use of brain prostheses, invisible colleges, and conceptual homunculi to help save us from overpopulating ourselves to extinction. To that end, a few years after his AAAS talk in Dallas, Calhoun began referring to himself as a ℞evolutionist, which he defined as "a new type of 'revolutionary,' where ℞ is the prescription for, or design of, evolution."[10]

11

THE BEAUTIFUL ONES IN UNIVERSE 25

In the mid-1960s, Stanford University population ecologist Paul Ehrlich hardly seemed bound for the late-night talk show circuit. Ehrlich had done his PhD work on the evolutionary history of butterflies but was best known in evolution and ecology circles for a landmark paper that he and Peter Raven published on the coevolution of butterflies and their plant hosts. Coevolution, a term Ehrlich and Raven made famous, occurs when changes to characteristics in species 1 create new selection pressures that lead to changes to traits in species 2, which in turn feedback and affect traits in species 1, and so on, back and forth.

The exquisite fit between the structure of flowers and the structures that their pollinators use to extract pollen and nectar from them has long been of interest to evolutionary biologists: it was front and center in Charles Darwin's book *The Various Contrivances by Which Orchids Are Fertilised by Insects*. On January 25, 1862, Darwin, part of a worldwide group of scientists who exchanged samples, received a box of plant samples from an orchid grower in Madagascar. That box contained an *Angraecum sesquipedale* orchid, a species with a very long tube (*sesquipedale* is Latin for "a foot and a half long") that leads to a pool of nectar at its far end. Darwin immediately wrote to a colleague, "I have just received . . . [an] astounding *Angraecum sesquipedalia* [*sic*] with a nectary a foot long. Good Heavens what insect can suck it." After mulling that question over, he predicted that there must be a moth species with a proboscis—an appendage for sucking nectar—long enough to pollinate this species of orchid, though such a pollinator had never been observed. In 1903, a candidate moth, *Xanthopan morganii praedicta* (*praedicta* is Latin for "predicted," in honor of Darwin's prediction), was discovered in Madagascar, and, at long last, in 1992 this species was photographed pollinating *A. sesquipedale*.

Looking at a vast array of butterflies and the plants they feed on in the tropics, what Ehrlich and Raven found, and what was later dubbed "escape and radiate" coevolution, was that natural selection in both butterflies and plants led to a spectacular diversity of insects and plants (their primary interest was in the former). Ehrlich and Raven's argument boiled down to this: many butterfly species attack and feed on roots, stems, leaves, flowers, and fruits. On occasion, mutations in plants that are being fed upon by butterflies lead to the production of a new chemical compound that protects the plants from the butterflies that feed on them but causes little harm to the plants themselves. That new lineage of plants, Ehrlich and Raven suggested, entered "a new adaptive zone" and might very well, over long periods of time, radiate and lead to new families of well-protected plants. There's a rub, though, and that is that eventually new mutations that allow butterflies to remain unharmed by the new chemical compound will appear in a butterfly species, which will then move into its own new adaptive zone and may seed new families of butterflies that can get around the new chemical defenses. Combine all this with the fact that this is happening not in one butterfly or plant species but many, that the same argument applies to *any* plant-insect system where the insect feeds on the plant (not just butterflies and their plant victim), and that any new adaptative zone will *eventually* be breached for the same reasons listed above, and, Ehrlich and Raven argue, you end up with the incredible diversity we observe in insects and plants we see.[1]

Ehrlich exploded onto the world stage with the publication of his 1968 book, *The Population Bomb*. Published by both the Sierra Club and

Ballantine Books, and with his wife Anne Ehrlich listed as "virtually a co-author," *The Population Bomb* pulled no punches in its claims that overpopulation was an existential threat: "*Population Control or Race to Oblivion?*" it asks on its cover. "Overpopulation is now the dominant problem in all our personal, national, and international planning. No one can do rational personal planning, nor can public policy be resolved in any area unless one first takes into account the population bomb." The population explosion, Ehrlich screamed from the hilltops, was a disease, and humanity needed urgent care. "The birth rate must be brought into balance with the death rate or mankind will breed itself into oblivion. We can no longer afford merely to treat the symptoms of the cancer of population growth," he told his readers. "The cancer itself must be cut out. Population control is the only answer." That control took many forms, including "population control at home," ideally by a new Department of Population and Environment creating a combination of penalties (each child after two would garner an additional tax burden, luxury taxes placed on cribs, expensive toys, etc.) and incentives ("responsibility prizes" given to married couples for every five years they go childless, subsidies for adoption, etc.) aimed at a reduction in the birth rate.

Steve Suomi, who decades later became the chief of the Laboratory of Comparative Ethology at the National Institute of Child Health and Human Development, and who eventually moved his laboratory into Calhoun's space at NIH in the mid-1980s, was an undergraduate at Stanford at the time Ehrlich was researching background data for *The Population Bomb*. Because Suomi was interested in both biology and psychology, one summer Ehrlich had him do a literature review on population growth in animals. Suomi had read Calhoun's 1962 *Scientific American* article when he was a senior in high school. "It absolutely fascinated me," he recalls now, some sixty years later. In the review he wrote for Ehrlich, Calhoun's Casey Barn rats were near the top of the list. Ultimately, Ehrlich opted to write little of nonhuman population dynamics in *The Population Bomb*. Still, Calhoun and the Casey Barn rats make a cameo, though not by name: "We know all too well that when rats . . . are overcrowded, the results are pronounced and usually unpleasant," Ehrlich wrote. "Social systems may break down, cannibalism may occur, breeding may cease altogether. The results do not bode well for human beings as they get more and more crowded." Ehrlich knew, though, that there was only so far one could go with that analogy: "Extrapolating from the behavior of rats to the behavior of human beings is much more risky than extrapolating from the physiology of rats to the physiology of human beings. Man's physical characteristics are much more rat-like than are his social systems." Calhoun never explicitly commented on that remark, but given his research interests at this point, one might reasonably assume that if he didn't believe that last sentence was outright wrong, at the very least, he'd have thought it an understatement of the value of rodent research for dealing with the human population explosion.

While *The Population Bomb* got attention from the media right from the start, it appeared on the national (and then international) radar after Ehrlich made his first (of eighteen) appearances on *The Tonight Show* with Johnny Carson in February of 1970, drawing thousands of letters from fans and securing his book a spot on the bestseller list. For the better part of the last fifty-plus years, demographers, sociologists, population ecologists, politicians, and more have been debating whether the specific predictions Ehrlich made were prophetic or ecological scare tactics by a radical liberal Stanford professor. Ehrlich told readers of *The Population Bomb* that his scenarios of what was to come were "just possibilities, not predictions." He continued, "We can be sure that none of them will come true as stated, but they describe the kinds of disasters that will occur as mankind slips into the famine decades." That said, Ehrlich *does* in fact use the language of specific prediction on occasion, including right at the start of the book, before that caveat (on page 72), and certainly some of his most dire predictions, such as "in the 1970's the world will undergo famines—hundreds of millions of people are going to starve to death in spite of any crash programs embarked upon now," were dead wrong. Writing forty years after the book's publication, Ehrlich admitted as much: while hundreds of millions may have starved in those forty years, they didn't starve in a decade. He attributed his error to the fact that he and others had underestimated our ability to increase the food supply.

Over time, Ehrlich came to regret some of the tactics he used in *The Population Bomb* but stands by the main message: "The biggest tactical error in *The Bomb* was the use of scenarios, stories designed to help one think about the future," he wrote in 2009. "In honesty, the scenarios were way off, especially in their timing (we underestimated the resilience of the world system). But they did deal with future issues that people in 1968 should have been thinking about—famines, plagues, water shortages." Indeed, one thing everyone agrees on is that *The Population Bomb* made human population growth a topic for dinner conversation around the world, drawing attention in a way that few books by scientists did.[2]

In the late 1960s, as *The Population Bomb* was being debated, Calhoun's work was not only being picked up by major newspapers like the *New York Times*, which covered the Casey Barn rats in "An Attack on Man the Aggressor: Scientists in Rush to Study the Pattern of Destruction," but popular books and movies were beginning to embed his work into the public's collective consciousness. The Casey Barn rats made their way into Tom Wolfe's 1968 book *The Pump House Gang*, published on the same day his *The Electric Kool-Aid Acid Test* hit the stands. Page after page of "Oh Rotten Gotham! Sliding Down into the Behavioral Sink," one of the chapters in *The Pump House Gang*, is devoted to Calhoun's studies. This was no happenstance: Wolfe learned of Calhoun's behavioral sinks from their mutual friend Edward Hall of proxemics fame.

"I just spent two days with Edward T. Hall, an anthropologist," the "Oh Rotten Gotham!" chapter opens, "watching thousands of my fellow New Yorkers short-circuiting themselves into hot little twitching death balls with jolts of their own adrenalin. Dr. Hall says it is overcrowding that does it . . . here they are, hyped up, turning bilious, . . .

sadistic, barren, batty, sloppy, hot-in-the-pants . . . leering, puling, numb—the usual in New York, in other words." Wolfe goes on to tell how "Dr. Hall has the theory that overcrowding has already thrown New York into a state of behavioral sink." He explains to his readers, "Behavioral sink is a term from ethology, which is the study of how animals relate to their environment. Among animals, the sink winds up with a 'population collapse' or 'massive die-off.' O rotten Gotham."

Wolfe knew just which animals too: "It got to be easy to look at New Yorkers as animals, especially looking down from some place like a balcony at Grand Central at the rush hour Friday afternoon. The floor was filled with the poor white humans, running around, dodging, blinking their eyes, making a sound like a pen full of starlings or rats or something." A few pages later, Calhoun and the Casey Barn rats take center stage: "In one major experiment, an ethologist named John Calhoun put some domesticated white Norway rats in a pen with four sections to it, connected by ramps," Wolfe begins. "He allowed them to reproduce until there were eighty rats . . . but did not let it get any more crowded . . . to the human eye, the pen did not even look especially crowded. But to the rats, it was crowded beyond endurance."

All the rat players in the Casey Barn make appearances in this chapter. "Two dominant male rats took over the two end sections, acquired harems. . . . [These] aristocrat rats grew bigger, sleeker, healthier, and more secure," writes Wolfe. "Slumming females from the harems had their adventures and then returned to a placid, healthy life. . . . Pregnant rats had trouble continuing pregnancy. The rate of miscarriages increased significantly. . . . Child-rearing became totally disorganized."

Naturally enough, Calhoun was pleased with the effect of Hall's influence on Wolfe's: "Ned [Edward Hall] and I share the view that social ideas become effective only after gaining coinage in common parlance," Calhoun wrote. "Although Wolfe used a considerable literary twist, many readers must have gotten the notion of traps we unknowingly can get into." Gonzo journalist Hunter Thompson certainly got that notion. He was enamored with the term *behavioral sink* and thought Wolfe himself had come up with it. "The term itself is a flat-out winner, no question about it," Thompson wrote Wolfe. "Every now and then I stumble on a word-jewel; they have a special dimension, like penetrating oil. Right? 'Behavioral Sink' is up in that league with my all-time, oft-used champ, the 'Atavistic Endeavor.'"[3]

Calhoun's ideas were also being discussed on the big screen. In 1969 fellow Space Cadet Ian McHarg not only had Calhoun lecture to his class at the University of Pennsylvania but he also used *Multiply and Subdue the Earth*, a documentary movie that he narrated for public television, as a means to highlight Calhoun's work on behavioral sinks and population growth in the Casey Barn rats.

Multiply and Subdue the Earth was largely a platform for McHarg to rail against humanity's disconnect with nature and our seemingly endless destruction of it. In the movie, in a strong Scottish accent, McHarg tells viewers, "We have never learned we are a part of nature. . . . Anthropocentric man seeks not purity with nature, but conquest . . . or even better, exploitation." McHarg's solution was "to crack our old value system based on ignorance and economics and adopt a new relation with nature based on ecology." He continued, "Let us ask nature herself what lands we can develop and which we must leave untouched." Still, no viewer could have reasonably surmised that a pitch for a new relation with nature based on ecology would be the trajectory of the film from what they saw in the opening minutes.

The first scene in *Multiply and Subdue* splices in older film of Calhoun's Casey Barn rats scurrying around one of their neighborhoods, with Calhoun narrating their tale. As Calhoun tells the audience that many "females [rats] never developed the ability to care for their young," the scene shifts from the Casey Barn to a young, giggling, quite tipsy couple in a bar. A few seconds later, viewers are back at the Casey Barn listening to Calhoun discuss "extremely deviant sexual behavior" in his rats, followed by another abrupt shift in scenery, this time to young men frequenting the sex district in New York's

Times Square, with a movie marquee advertising the flicks playing there, including *Love Now, Pay Later* and *The Naked Temptress*. Then it's back to Calhoun and his rats, this time to hear him describe the phlegmatic, subordinate, socially withdrawn rats in the barn. As he does this, the scene turns to throngs of New Yorkers, seemingly oblivious to those around them, emerging onto a street from a subway station, at which point Calhoun ends his role in the movie by warning viewers that his "totally withdrawn [rats]," all unaware "of the presence of any other even though packed close together . . . might have application to the human scene."[4]

All this coverage focused on Calhoun's work in the Casey Barn rats—largely because his only public presentation so far about the mouse universes (his 1968 AAAS presentation) outlined only the most preliminary of results. But now more data from the mice in Study 102 were flowing in. Results that were strikingly similar to what he had seen in his rats. While Calhoun was interested in population growth and behavior in all the mouse universes—two-, four-, eight-, and sixteen-celled—it was what was going on in the sixteen-cell universe that was proving most interesting to him. For one thing, in this universe—which Calhoun labeled Universe 25—as the population grew and grew, mice began aggregating at the feeders, and by pairing feeding with social interactions, they began sending the population down the behavioral sink, though at this point, not completely down the drain. Dominant males fought for control of their space and their harems of females, and as a result, many male mice had dried blood stains on their fur. Again, as with the rats in the Casey Barn, there were the somnambulists, withdrawn, ignoring all social interaction, including mating, focusing solely on feeding and grooming.

These somnambulist mice seemed even more isolated than their rat counterparts years back. So striking was their shiny white fur—the result of obsessive self-grooming—that Calhoun began referring to them as "the beautiful ones." They were, as Calhoun saw it, "blobs of protoplasm, physically healthy, but socially sterile"—mice that were "frozen in a child-like trance" and "with little capacity to engage in those complex behaviors essential for the survival of the species." Among the Beautiful Ones, Calhoun made special note of two personality types. There were some male Beautiful Ones who would "line up, on the low elevation bars separating the public floor space of adjoining cells. Most of these would face the same direction." Calhoun called them "barflies." Then there were the "pied piper" Beautiful Ones. Always female, pied pipers "would collect about any strange object, including [a human] investigator, abruptly appearing in a universe. If the investigator began moving slowly about the universe floor, the mice aggregated about his feet would follow closely." Calhoun felt for the pied pipers: "These poor beautiful mice never learn," he told a *Washington Post* reporter. "Each day they will follow human feet about as if they have never seen them before." And to make them even more pitiable, Calhoun noted that the pied pipers were voiceless. "They are deathly quiet. Lost are the plaintive squeaks of recognition, the higher squeaks of inquisitive anxiety, the shrill squeaks accompanying intense social involvement. A parlor of silence hangs over the flaccid following mass."

One of Calhoun's coinvestigators on Universe 25 was fellow NIH scientist Julius Axelrod. A world-renowned biochemist, Axelrod shared a Nobel Prize for his work on the release and update of certain key chemicals, particularly in the brain, in response to danger and aggression. When Axelrod measured the rate that chemicals were transformed into adrenaline in Calhoun's mice, he found, particularly in the later years of Universe 25, that Beautiful Ones had low levels of adrenaline. Calhoun and Axelrod interpreted this to mean that even when population numbers were rising, the Beautiful Ones, trapped in their own asocial niche, remained unstressed.[5]

As mouse Universe 25 marched on, and as plans for rat culture experiments were being drawn up, Calhoun continued to write and speak about bridging the gap between population explosions in nonhumans and humans. In "Design for Mammalian Living," a paper that he published in *Architectural Association Quarterly*, the journal of the British Architectural Association, Calhoun outlined his work on the Towson enclosure, the Casey Barn, the cooperative and disoperative rats, and the mouse universes. In an article he wrote for an offbeat journal titled *Man-Environment Systems*, Calhoun referred to himself as part of the metaenvironmentalist movement, but one with a different perspective than was typically associated with that movement. Other metaenvironmentalists were "concerned with bricks and mortar, highways and subways, the purity of air and water," but for Calhoun, these "merely represent[ed] necessary means or tactics to fulfill broader strategy." While Calhoun saw those concerns as valid, he wanted to work from first principles.

Those principles involved incorporating evolutionary ideas, focusing on "enhancing creativity," "increasing valuing with respect both to recognizing and adapting new images and to relinquishing values no longer optimal," and the development of a global brain, with "networks of communication whose nodes are both diverse and numerous." In some notes Calhoun jotted down after attending a Conservation Foundation meeting on "Population and the Environment," Calhoun casts humanity as a global zoo. Adaptation in that zoo occurred "when channels of contact develop between groups whose partial isolation has permitted them to differentially specialize from one another. Neither complete isolation nor complete integration has proved evolutionarily most profitable for survival." It was instead "the promotion of the twin 'explosions' of enhancement of creativity and enhancement of ability to change our values" that was most important. What we need to do, Calhoun proposed, was to "design a zoo where the contained cells become more diverse, but with each cell, whether spatial or ideational, interconnected or linked to several other cells of different kind."[6]

In 1970 Calhoun's work was making its way into books written for public consumption, including Robert Ardrey's *The Social Contract: A Personal Inquiry into the Evolutionary Sources of Order and Disorder*. Ardrey, a playwright and screenwriter who had studied under Thornton Wilder and had been nominated for an Academy Award for his screen writing, had a long-standing interest in biology and anthropology, which had laid dormant for a time, until he went to Africa and met with paleontologist Raymond Dart and examined the *Australopithecus africanus* fossils Dart had discovered. Ardrey had made passing reference to Calhoun's work in his 1966 book, *The Territorial Imperative*, but devoted much more print space to Calhoun's rats and mice in *The Social Contract*.

Dedicated to "the memory of Jean-Jacques Rosseau," the opening pages of Ardrey's *Social Contract* map out what lay ahead for readers: "A society is a group of unequal beings organized to meet common needs. In any sexually reproducing species, equality of individuals is a natural impossibility," Ardrey claims, because of genetic differences that are part and parcel of sexual reproduction. "Inequality must therefore be regarded as the first law of social materials," he writes, "whether in human or other societies." But no need to despair, says Ardrey, as "equality of opportunity must be regarded among vertebrate species as the second law. . . . Every vertebrate born, excepting only in a few rare species, is granted equal opportunity to display his genius or to make a fool out of himself."

Ardrey argues that understanding nonhuman societies can help us create a better society for ourselves: "The just society . . . is one in which sufficient order protects members, whatever their diverse endowments, and sufficient disorder provides every individual with full opportunity to develop his genetic endowment, whatever that may be." But his was a social contract with commandments: "It is this balance of order and disorder . . . that I think of as the social contract . . . [that] it is a biological command[ment] will become evident, I believe, as we inquire among the species." Among those species were rats, particularly those of the Casey Barn.

Ardrey was enamored with Calhoun and his rats. The rat work, he warned his readers, had "horrifying human implications." As for Calhoun himself, he was "a maverick's maverick in the field of psychology," Ardrey wrote. "Physically slight, temperamentally elusive in the sense that elves are hard to get hold of, Calhoun is blessed with the capacity of slipping through the formidable fences of American psychology."

Ardrey discussed the Towson enclosure work and the work on the Casey Barn rats, which he described as one of the "more ominous experiments" ever done on population growth. As in the subtitle of his book, Ardrey saw order in the neighborhoods with dominant rats and disorder and a behavioral sink in the other rat neighborhoods in the Casey Barn experiment.

From ecology to anthropology to sociology and more, Ardrey's *Social Contract* was largely panned as sloppy, albeit well-written, pop science. Ardrey didn't help himself on this front with statements like "The three sciences central to human understanding—psychology, anthropology, and sociology—successfully and continually lie to themselves, lie to each other, lie to their students, and lie to the public at large, [and] must constitute a paramount wonder of a scientific century. Were their condition generally known, they would be classified as public drunks."

A review in the *American Scientist* bemoaned that *The Social Contract*'s "lack of coherence will bother the analytical reader, and the bristling and biased hostility cannot but dismay the scientist." Ardrey, the reviewer went on to say, was "an excellent raconteur," but his book was "misanthropic." Sociologists were a bit kinder: writing of "jewels . . . mixed with drek," in *The Social Contract*, while anthropologists warned that "inaccuracy or intellectual sloppiness which cannot easily be detected by the general reader seeking an insight into the evolutionary basis of the human condition, is especially deplorable." And Ardrey's book was guilty of this crime over and over. But all this was within the walls of the academy: as far as getting Calhoun's mice and rats to the public, *The Social Contract* was reviewed, sometimes kindly, sometimes not, in the *New York Times, The Observer*, and other major newspapers and magazines. As for Calhoun himself, the publisher sent him a draft of *The Territorial Imperative* hoping for a blurb or book review but got neither. In his handwritten notes on the book draft, Calhoun wrote, "Ardrey is a child, continually exploring, continually fixating on some transient image of reality, and just as continually, laughing at the absurd views of one of his former selves."

Calhoun's work was also a subject of discussion in Ronald M. Linton's 1970 book, *Terracide: America's Destruction of Her Living Environment*. Linton, a former director of Economic Utilization Policy in the Department of Defense and national coordinator of the National Urban Coalition, told his readers of the Casey Barn rats and their implications for the population explosion. Linton knew Calhoun through mutual connections in the federal government hierarchy and sent him a draft of the chapter in *Terracide* that dealt with Calhoun's work, but that package got lost in the NIH mail system for so long that Calhoun did not have a chance to send back comments. In a more bizarre vein, one of Calhoun's key ideas, though not his name or his rats or mice, was featured in *Insect Fear: Tales from the Behavioral Sink*, a 1970 X-rated, "for Adult Intellectuals Only" comic book set in an urban cesspool of a city, stocked with a combination of humans and humanoid creatures.[7]

When a visitor to his lab asked Calhoun whether he used inductive reasoning—roughly speaking, moving from the specific to the general—or deductive reasoning—again, roughly speaking, moving from the universal to the specific—in his research, he replied, "Neither: rather we approach our problems poetically. By 'poetic' I mean most nearly what Thomas Kuhn means by the term 'revolutionary science': or is employed by the word 'serendipity' coined by Horace Walpole." An unusual response from an NIMH researcher for sure, but Calhoun wasn't through yet: "The tactic is to describe some set of circumstances out of which new insights evolve and to end with the comment: 'you now see what I mean!' I too must follow this tactic and feel no self-deprecation from admitting to not fully understanding the origin of the creative act."

Putting aside the origin of the creative act that led to Study 102 (including Universe 25), by 1970 Calhoun was beginning to get a much better understanding of population dynamics in all the universes in that study. While larger universes did have smaller populations early on, that changed as time passed, and sixteen-celled Universe 25 had more than two thousand mice: by far, the largest population in any of the universes. More generally, populations in all universes tended to go through four distinct phases. Early on, colonizers learned the lay of the land, followed by a period of rapid population growth in which mice competed to establish territories and most litters were healthy and survived. Next came a period of inhibited growth and hints of the emergence of a behavioral sink. Younger mice experienced little in the way of stress, but as they matured, "except for a few who were able to replace their older associates, all were rejected, they found no opportunity for expressing their capacity for social involvement," and they were chased from already existing groups in the universe. They simply lived in a universe that was starting to get overcrowded, and they were shunned from mouse society. The males became very aggressive with others who had been shunned, but not with those in the standing order, and females became poor mothers. It was the offspring of these individuals who were the Beautiful Ones, as they "never had the opportunity to develop those critical aggressive, sexual, and maternal behaviors requisite to the survival of the species."

By 1970 Calhoun's mouse universes had moved to a period of relative stability, with populations leveling off. At this point, more than 90 percent of the mice in Universe 25 were Beautiful Ones: Calhoun told one interviewer that "between twenty-two hundred and twenty-five hundred mice lived in that universe," quickly qualifying his comments, "but there are really only about a hundred and fifty mice in here. Real mice that is, mice who are in the social sphere." Showing one of the Beautiful Ones to the interviewers, he simply noted, "he's a total nothing."[8]

During the summer of 1970, *Newsweek* reporter Stewart Alsop paid a visit to Calhoun's universes. "It's a lovely day—much too lovely to spend in an office," he wrote in his August 17 *Newsweek* piece. "In fact, it seemed a perfect day to visit Dr. John Calhoun's mousery."

Alsop paints Calhoun as "a smallish, cheerful man with bright blue eyes, a goatee and mouse-colored hair that looks as though it had never been combed . . . [a man] whose chief interest, he soon made clear, is not in mice but in men." At this point, Alsop proceeds to tell *Newsweek* readers of Calhoun's fears regarding the human population explosion. "All this was familiar, and because familiar, not really disturbing," Alsop added, "but then Dr. Calhoun led me upstairs to his mousery and gave me a horrible glimpse of what may be, after all, just around time's corner."

Calhoun introduced Alsop to Universe 25 when that universe was home to more than two thousand mice and showing signs of a behavioral sink. Mice moving about in a frenzy, and the odor of thousands of them doing so, was a bit much for Alsop, who had "a strong impulse to get out again, into the sunlight." But when Calhoun began explaining what was happening in his universe, Alsop "stayed, fascinated and appalled."

Calhoun pointed out the dominant mice and told Alsop how they had set up their fiefdoms, which, Alsop relayed to his readers, was a mouse polity where "the rodential Bourgeoisie established themselves with their consorts in the higher nesting boxes, nearest the food and water." And then Calhoun not only pointed out the Beautiful Ones—he also reached down into Universe 25, scooped up two of the finest of them, shiny fur glimmering, and, along with four other males he grabbed, handed each of the mice to Alsop. "The difference was obvious," Alsop wrote. "The proles were scruffy and chewed up—one had lost half its tail—while the beautiful ones were sleek, unharmed and utterly passive." Alsop then turned to Calhoun and asked, "Aren't we maybe seeing the phenomenon of the beautiful ones . . . in the dropout, drug culture?" Calhoun replied that the science was not in place to answer that question, but to Alsop's surprise, "he did not think the question ridiculous."

Alsop's article ended with an ominous question of things to come: "This generation of the young, unlike their elders, will live to see Dr. Calhoun's upper [population] threshold reached. Is it possible that when the threshold is reached, population growth will be ended, not by birth control or the bomb, but by the mysterious and terrible process that ended all reproduction in Dr. Calhoun's horrible mousery?"

The *Newsweek* article generated tremendous attention, leading to dozens of invitations for Calhoun to speak at universities all over the country. Calhoun declined the majority of these with pro forma, but diplomatic, responses, but he was particularly disappointed that time commitments forced him to decline an invitation from the students of Virginia's Fairfield High School student government, who asked him to speak. "I was fascinated by [the] article in the *Washington Post*," wrote the president of the student government. "The problem of overpopulation is such a relevant one. . . . We would be very pleased to have you speak to our student body on your studies and opinions about the effects and control of overpopulation. Quite a few students have mentioned your article to me." Calhoun was touched by the invitation: "It causes me considerable distress to say to you that I will be unable to participate in your international week program," he replied. "The pressure of keeping our studies going keeps my nose tied to the grindstone for some time in the future."[9]

12

THE (REAL?) RATS OF NIMH

Alsop's *Newsweek* piece on Calhoun's rats at the NIH field station read like a dystopian science fiction story. Of course, it was anything but fiction. The 1971 children's classic novel *Mrs. Frisby and the Rats of NIMH*, on the other hand, was, and there was nothing dystopian about those rats (or mice). The book jacket of the fanciful saga read: "Nothing [she] had heard of the rats was as strange as the truths she discovered. . . . All had been imprisoned for several years in a laboratory known as NIMH."

Writing under the pseudonym Robert C. O'Brien, Robert Conly begins his story with the tale of a widowed mouse—Mrs. Frisby—who is desperately trying to save her son Timothy, who has fallen ill with pneumonia. Mrs. Frisby doesn't have the luxury of time, as her home in a cinderblock on Farmer Fitzgibbon's land is threatened with imminent destruction when the farmer's plow next comes through for spring planting. Throwing caution to the wind, Mrs. Frisby visits a wise owl. The owl tells her of a group of rats living under a rosebush who might be able to save Timothy.

These rosebush rats, led by one-eyed Nicodemus, had once been imprisoned and subject to experimentation at a place called NIMH. As a result of that experiment, which involved various chemical injections, Nicodemus and the others at the rosebush were rat geniuses, capable of reading, writing, and building and operating machinery.

Nicodemus and the others agree to help Mrs. Frisby and eventually explain why: it turns out that her deceased husband had also been imprisoned at NIMH, and that he and other mice there had played a pivotal role in helping Nicodemus and his compatriot rats escape from their cages, freeing them from their servitude to humans. In time, after wonderful twists and turns, Nicodemus and the other rats of NIMH save Mrs. Frisby and her family by moving them to safety.

Mrs. Frisby and the Rats of NIMH went on to win a Newbery Medal in 1972, and a decade later, in the summer of 1982, an animated movie adaptation called *The Secret of NIMH*, was appearing in theaters around the country. Three weeks after the movie was released, Sandy Rovner wrote a *Washington Post* piece titled, "Rats! The Real Secret of NIMH," which suggested *Mrs. Frisby and the Rats of NIMH* might have been inspired by Calhoun's work at the NIH field station. Rovner interviewed Calhoun for the article and wrote, "He remembers the late O'Brien, the book's author, visiting the facility in the late '60s or early '70s. In fact, Calhoun believes that Mrs. Frisby's name came from the blue Frisbee he kept hanging on his door 'to help when things got too stressful for us.'" There's little reason to question Calhoun's memory here: Conly lived near Washington, DC, and had been a reporter for *National Geographic*, and his daughter, Jane, who later wrote a series of follow-up *Rats of NIMH* books, says her father did visit NIMH around that time, though she doesn't mention Calhoun's lab per se.

Rovner seems ambivalent about whether *Mrs. Frisby and the Rats of NIMH* had any deep ties to Calhoun's work: "In the story, some of their exploits do seem to reflect some of the rat-cultural happenings in Calhoun's overcrowded rat population—leadership rivalries, for example," Rovner tells readers. "But Calhoun's rats weren't injected with anything. They were just crowded." All of which was true, but Calhoun wasn't ambivalent in the slightest. Above and beyond Frisby/Frisbee, Calhoun was quite sure that Conly had, at the very least, been inspired by his visit to the lab, and had perhaps read some of

Calhoun's journal articles or media write-ups of them. Sometime after Calhoun's death, his daughter Catherine found an eleven-page handwritten document buried in one of her father's books. In that document, which was written at some point after 1982 (when the film adaptation came out), Calhoun systemically links his work to Conly's book, going as far as to cite page numbers in *Mrs. Frisby and the Rats of NIMH* alongside his notes on how material on that page connected to his own work. Many of those connections involved the layout of Calhoun's rat and mouse universes (spiral staircases and the like) as compared to the world inhabited by the rats and mice in *Mrs. Frisby and the Rats of NIMH*, but there was more to it than that, at least as Calhoun saw it.

When Mrs. Frisby first encounters the rats, there were a dozen of them marching in the field. No number was more critical to Calhoun's invisible college idea than twelve, not because it appears so often in our culture—twelve months, twelve-person juries, twelve days of Christmas, and so on—but rather in part because of the experimental work he had done on the rats and mice, and in part because of the mathematical models that he built of optimal group size. The cable the rats were moving in the field struck Calhoun as reminiscent of the levers that rats had to push in his STAW cooperation experiments, and Conly's rats displayed all sorts of cooperation moving the Frisby home to a safe locale. Conly's visit had been so long ago Calhoun couldn't recall if he had mentioned the global colleges or shown Conly the STAW device, but it was possible—or perhaps Conly had read about them. Besides helping Mrs. Frisby, the rats were working on "The Plan of the Rats of NIMH." That plan sounded vaguely familiar, and in his handwritten notes, Calhoun asked himself, "Did O'Brien get the London paper on Environmental Control over Four Paths of Mammalian Evolution?"—by which he meant the "Design for Mammalian Living" article he had written for the British Architectural Association.

Neither Conly, Robert nor Jane, ever commented on what role, if any, Robert's visit to Calhoun's lab played in the development of *Mrs. Frisby and the Rats of NIMH*. At the very least, then, readers of the *Washington Post* were left with a sense that Calhoun's rats *might* have played a role, and Calhoun was certain they did. Indeed, although nothing ever materialized from it, Calhoun approached NBC television about filming a documentary in his lab that would make it clear the role his work played in spurring on *Mrs. Frisby and the Rats of NIMH*.[1]

It was one thing for Calhoun's work to be discussed among academics, as well as in newspapers, books, and films, but it was another matter altogether for it to be enshrined in the *Congressional Record*. But that's exactly what happened on April 1, 1971, when Senator Robert Packwood, a Republican from Oregon, stood before his colleagues at the Capitol, pleading with them to take the population explosion seriously. "[A] critical aspect of the population problem is being studied by Dr. John Calhoun of the National Institute of Mental Health," Packwood told his fellow senators. "The pressures of overpopulation have resulted in extreme antisocial behavior and eventually failure to continue reproducing." He next informed his august colleagues that "Dr. Calhoun is careful not to transfer the results of his experiments with mice into the human experience directly, but certain lessons can and should be learned as our finite earth looks forward to its 4 billion people." Packwood went on to ask "unanimous consent that a write up of Doctor Calhoun's work(s) . . . be printed in the *Record*." That unanimous consent was granted, and so it was that Calhoun's *Scientific American* article about the Casey Barn rats and two of his other papers were entered into the *Congressional Record*.

Packwood was inspired to act when he did by a *Washington Post* story published three days earlier (and also entered in the *Congressional Record*). In that article, "Of Mice and Men," reporter Tom Huth told his readers of mouse life in Universe 25: "Most of them are withdrawn," he wrote, "uncomplaining, uninvolved, without aggression.

And they are without sex so they are dying." The "and Men" part of the *Washington Post* headline had Huth writing of Calhoun's ideas on appreciative systems, conceptual revolutions, and a global brain.

Huth described Calhoun as "a psychologist, philosopher, economist, mathematician and sociologist at various moments," and pressed him about one concern about his work in the larger scientific community. Huth told the readers of the *Washington Post* that "to conventional science there is a madness to Calhoun's methodology. He develops and publishes theory without first presenting full documented findings in scientific journals. He draws conclusions and exposes them to public attention before his experiments are completed." To be fair, while Calhoun was certainly not averse to publicity in venues like the *Washington Post*, the *New York Times*, and the like, it was usually reporters (like Huth) who reached out to him, rather than the other way around. But Huth's criticism was not completely unfounded. By this time, though he had many experiments running, Calhoun had largely, but not completely, stopped publishing his experimental results on rats, mice, and overpopulation in science journals. "I had ceased publishing in the most common way," Calhoun wrote elsewhere. "That is, I had quit submitting unsolicited articles for publication in journals. Instead, I just tried to keep pace with invitations to publish in collected works, such as symposia." Calhoun's defense of this, he told Huth and his readers, was that there just wasn't time to wait before impending human population explosion caused irrevocable harm, and "so sometimes we make errors in what we say," though he provides no details on what those errors were.[2]

About the time Senator Packwood was discussing his work in the halls of Congress, Calhoun, Paul McLean (now in charge of the group Calhoun was part of), and others had come together to create an interdisciplinary Laboratory of Brain Evolution and Behavior at the NIH field station. The powers-that-be at NIH funded the construction of two new buildings for this new entity, and Calhoun had grand aspirations for what could be done there: it would be a place where a "holistic approach to the interaction between the social and physical environments lead to the projection of insights gained onto the human scene and man's own spaceship earth bound environment."[3]

Not long after Calhoun, his mice, and his rats moved into their new home, the nearly nine thousand square feet of space in Building 112, an "unimposing metal prefab shell with its interior walls painted a rather bureaucratic orchid color," the media swarmed in. Nina Laserson, a writer for *Innovation* magazine, paid Calhoun and his rodents a visit in Building 112 in late 1971. She titled her resultant story "It's Not Every Day You Walk into a Laboratory Whose Mission Is to Save the World." The article painted Calhoun's work in a positive light, and Laserson portrayed Calhoun's lab as one whose mission was to save the world, at least from the population explosion. But the piece ended by planting the seeds of doubt that work with rats and mice using government funds could provide answers about human population growth in time to save us. After describing Senator Packwood taking to the floor of the Senate and telling his colleagues of Calhoun's work, Laserson wrote, "So people are sitting up and taking notice. But it took three months to acquire the orchid paint [for the walls of the lab] through the intricate mechanisms of government, how long is it really going to take to change the world?"

Japan's oldest newspaper, the *Mainichi Shimbun*, followed suit and had Yoichi Shimizu, one of their reporters stationed in Bethesda, visit Calhoun, who told him all about Universe 25: "The inhabitants of [this] universe moved on straight towards destruction," Calhoun told Shimizu. "They have lost every incentive including that of reproduction and they merely exist to sleep and eat. Their social [lives] have long ceased to exist . . . you can assume that they represent the human being on the limited space called the earth." The *Mainichi Shimbun* article is largely Shimizu conveying what Calhoun told him verbatim, but toward the end, Shimizu included comments from "authorities of various fields." A researcher from the Tokyo Ueno Zoo noted that after reading about Calhoun's work, he "suspect[ed] something similar—though perhaps in a different degree—[was] happening in the past several decades in large cities such as Tokyo or New York. Unexplained irritability, aggressiveness and murder without motives may be examples." A professor of Indian philosophy and Buddhism at Toyo University in Japan saw the breakdown of rodent societies in a more religious way: "Buddhism preaches against the vice of greed," he said, "showing examples with snakes and natural disasters striking mankind. I believe this experiment substantiates Buddhist teachings once again."[4]

Other newspapers and magazines that didn't have their reporters come and spend a day in Calhoun's mousery and rattery also wrote about both. *Der Spiegel*, a German weekly news magazine with the highest circulation of any such magazine in Europe, ran a piece similar to Alsop's *Newsweek* piece, as did Sweden's leading daily newspaper, *Svenska Dagbladet*, in their article titled "Of Mice and Men." And Canada's *Globe and Mail* ran a story on a lecture Calhoun had given about both Universe 25 and human population growth at a recent UNESCO (United Nations Educational, Scientific and Cultural Organization) meeting. In the United States, *Time* magazine published "Population Explosion: Is Man Really Doomed?" which warned readers that "if the progression continues, it is widely and gloomily predicted by the spiritual heirs of Thomas Malthus, there will be 7 billion people standing in line for their rations in the year 2000. By 2050, perhaps 30 billion will be fighting like animals for a share of the once-green earth." Even if it never got quite that bad, reporter Otto Friedrich told his readers, we "may still face a psychic fate similar to the one that befell Dr. John Calhoun's white mice . . . hardly alive."[5]

In his June 1971 NIMH progress report, Calhoun wrote that he was keeping Universe 25 up and running "in an effort to determine whether, in (a) population exposed to severe overcrowding, resultant physiological and behavioral pathology precludes any possibility of reproductive or social recovery." By this point, the thousands of mice in Universe 25 made it impossible for Calhoun and his team to gather detailed behavioral observations on each mouse. Instead, with his proclivity for groups of twelve, Calhoun selected eight groups of twelve male mice and eight groups of twelve female mice and used the combination of colors painted on their fur to identify individuals and record eight thousand behaviors by these 192 mice, mostly by observers taking handwritten notes but on occasion by video or time-lapse photography.

On March 1, 1970, 560 days after its initiation, 2,200 mice called Universe 25 home. From that day the population began declining to the 1,400 that were present when Calhoun wrote his 1971 progress report. "During this decline," he noted, "no young survive to weaning and the incidence of pregnancy itself declined to zero." When interviewed by Maya Pines, an NIH reporter writing a story on Calhoun's work for NIMH's *Mental Health Program Reports*, Calhoun said of the near total collapse of the Universe 25 society, "All capacities for developing and maintaining social bonds are disappearing with the last group of animals born. These are mostly beautiful ones unstressed by normal tendencies for sex and aggression and they cannot learn. They're all washed out."

Calhoun's plan was to let Universe 25 run its course, and, he predicted, eventually extinguish itself. As with the rats at the Casey Barn a decade earlier, he and his colleague, this time Halsey Marsden, removed two dozen males and two dozen females and introduced them into "uncrowded environments" to see whether, after being overcrowded their whole lives, they might be able to "recoup normal behaviors." The initial results were not promising: the mice "recovered only minimal amounts of aggression, sexuality, and social organization." That was true no matter what combination of mice was placed in a new, uncrowded abode: Marsden created groups of only dominant individuals, only Beautiful Ones, and combinations of old and young mice but reported, "The only animals that have been 'cured' so far are a handful that were put in a sort of solitary confinement for two months (the equivalent of six years for a human being) after which they seem to be able to start life all over again, as if reborn."[6]

Calhoun's progress report also let the higher-ups at NIMH know that his rat cooperation work was getting closer to recommencing: "Our objective here is to develop a 'physics' of social behavior and social organization," he told them. His goal was no less than to use the STAW devices to examine "the theoretical origins of human culture." The technology being developed for these studies was impressive for the early 1970s. Each rat would have a "passive resonator," a glass-encapsulated coil implanted under the skin of its belly. Resonators were all tuned slightly differently, so that when a rat went under any of the electromagnetic portals placed throughout its environment, information on that individual was transmitted back to a computer. Others at the NIH field station were calling the setup that Calhoun was building a "socioenvironmeter" and planning to build others for their own work. Calhoun's socioenvironmeter could be set to allow only certain rats (dominant, subordinate, male, female, etc.) to pass through a portal. It was a powerful new tool for studying behavior and population dynamics and one that Calhoun knew might raise some eyebrows: "This technology has an aura of Orwellian Big Brotherism," he wrote. "Like any technology it can be misused." Seriously misused. Elsewhere, he wrote that he had already gone through two rounds of miniaturization with the passive resonator implants in his rats and that "carrying such a miniaturization only two or three steps more will bring it down to a point where an implant could be injected under a person's skin without his being aware of the item injected." He continued, "I am still enough of a coward to worry about possible uses that might be made of a technology developed solely for research purposes." Still, at least with respect to the rats, he assured readers, "We believe our use of it is highly positive one. We hope to clarify the principles whereby the enhancement of individual capacity overcomes the deleterious consequences of increased density."[7]

"Experts Get Look at—and Smell of—DC Jail" read the front-page headline of the *Washington Post*'s Sunday's City Life section on October 17, 1971. Above the headline were photos of the housing and mess hall at the prison, and below it sat a photo of the four experts the headline referred to: Paul Chernoff, from the parole board; Ronald Goldberg, an attorney representing some of the inmates; Karl Menninger, a psychiatrist with an expertise in penology who had recently published his book *The Crime of Punishment*; and . . . John Calhoun.

The DC jail in question was a temporary holding block. It was built to accommodate 663 prisoners but was housing 1,300 at the time, including 400 in one maximum-security block, where solitary confinement meant each prisoner sat in his own six-foot-by-five-foot cell twenty-two hours a day. The inmates, working with the Public Defender's Office, charged that the jail was "overcrowded, unsanitary and contribute[d] to 'cruel and unusual punishment,'" and they had been granted an order from US District Court judge William Bryant requiring a panel of experts be sent to the jail to inspect the conditions there.

In the *Washington Post* article, Calhoun was described as a "Mental Health research psychol[o]gist who has done overpopulation studies purporting to prove that overcrowded colonies of mice suffer deterioration." The conditions in the jail were appalling, in part because of overcrowding. Calhoun had been sent in to comment on how his work on population dynamics and crowding might shed light on possible solutions. In a telling demonstration of just how deeply Calhoun believed his work on rodents and the population explosion was tied to overpopulation and crowding in humans, he accepted the invitation. Hoping he might get some juicy quotes from Calhoun, William Claiborne, the reporter covering the story, was clearly disappointed, noting in his last mention of Calhoun that "he refused to relate his mice experiments to the jail population." Not because he didn't think there was a relationship, but because Calhoun was "saving his conclusions for testimony during a hearing on the lawsuit."

The visit to the jail had a profound effect on Calhoun: the setting was, he thought, designed to convey a denigrating message: "you are small and insignificant, a uniform component of an overpowering authority." Prisoners in solitary confinement, Calhoun wrote in an unpublished document titled "Remarks on a Visit to the D.C. Jail," had less space than that mandated space for dogs used in scientific research—"a situation in which man is treated less than a dog with regard to spatial requirements."

A few years later, in *Campbell v. McGruder*, the US District Court for the District of Columbia ruled in favor of the inmates. While we don't have a record of Calhoun's testimony in the trial, we do know this much: his work played a role, however small, in reforming the DC penal system. And that effect was long lasting. In his 1988 book, *Prison Crowding: A Psychological Perspective*, Paul Paulus wrote of both the Casey Barn rats and the mice of Universe 25, and in his 2012 book, *The Environmental Psychology of Prisons and Jails: Creating Humane Spaces in Secure Settings*, Richard Wener noted that "if there is a common human condition that can approximate the living conditions of rats in Calhoun's famous behavioral sink . . . , it is most likely to be found in prisons."

As Calhoun was visiting prisons in Washington, DC, Jonathan Freedman, who was a young professor of environmental psychology at Stanford, was employing Calhoun's rodents in his studies on density and crowding in young adults. "As soon as you looked up 'crowding' in those days," Freedman says, "you got Calhoun. . . . [Also,] most psychology textbooks, particularly social psychology textbooks, quoted Calhoun favorably . . . even introductory textbooks quoted Calhoun."

Among other things, Freedman's laboratory work involved placing high school and university students into different-sized rooms, in different densities, and then giving the students a suite of cognitive tests. Before he gets into the details of all that work in his 1975 book

Crowding and Behavior, Freedman opens with chapters on "Colonies, Swarms and Herds" (chapter 2) and "From Mice to Men" (chapter 4), with explicit mentions of Calhoun's Beautiful Ones.

More and more often where there were discussions of human population dynamics, crowding, and behavior, there were discussions—often lengthy ones—of Calhoun's rats and mice, including an eight-page paper that appeared in *Science*. There, the authors began their article—"Population Density and Pathology: What Are the Relations for Man?"—with "a review of some [animal] studies, noting the implications of possible animal-human similarities." The first, and lengthiest, example was the rats of the Casey Barn, including a mention of Calhoun's somnambulists.[8]

The Smithsonian Institution, too, was interested in Calhoun's studies. When Calhoun published "The Lemmings' Periodic Journeys Are Not Unique," his first of two articles in the *Smithsonian* magazine, the magazine was only a year old. But readers of *Smithsonian* were already well acquainted with both Calhoun and his work, as the maiden issue of the magazine, in 1970, included Frank Sartwell's lengthy piece, "The Small Satanic Worlds of John B. Calhoun." "Each of us has his own universe," Sartwell opened the article. "John B. Calhoun, research psychologist at the National Institute of Mental Health (NIMH) has several . . . each of them looks like a pet shop in hell, aswarm with mice and rats." Sartwell reviewed Calhoun's work from his days in graduate school to the Towson enclosure through to Universe 25, always casting a dark picture of Calhoun's rodents and man's future. Very dark. "Dr. Calhoun sees us becoming so overcrowded," Sartwell wrote, "that without realizing the forces at work we will be driven to . . . cannibalism and extreme withdrawal amounting to a denial of life." Sartwell did not attribute those exact words to Calhoun, and so, on the surface, they might just represent some journalistic liberty on his part. This turns out not to be the case: *Smithsonian* sent Calhoun a prepublication draft of the article, and while Calhoun marked up the draft in many places, he left Sartwell's sentence on cannibalism and a denial of life intact, suggesting he saw that as one scenario for our fate should we fail to control population growth. But Calhoun's other writings from this period, which focused on his compassionate revolution, suggest that he thought of this *Smithsonian* description as a worst-case scenario.

Sartwell wrote that in Calhoun's universe, "Males are painted as gaily as a totem pole to identify their characteristics. Alternating stripes of red, yellow, green, and black brand them as aggressors, hangers-back, former or present malingerers or withdrawn animals." As for Calhoun, he "does not fit the picture of a prophet of doom," according to Sartwell. "A suburbanite, whose 15-year-old daughter plays the oboe, he sports a snappy gray goatee. Although full of doleful predictions, he seems as chipper as an unthinking sparrow when he darts around one of his sheet-metal universes." Yet this was a man, the readers of *Smithsonian* learned, who was certain his work with rodents could help us defuse the human population bomb. Sartwell ended his article with this: "As a prophet, John Calhoun is not entirely without Honor . . . He cries the warning . . . others listen—but not those who are breeding the world toward the doom he expects."

Calhoun's "The Lemmings' Periodic Journeys Are Not Unique" article was ostensibly about the famous Norwegian rodents that migrate to the sea and perish, but, as readers discovered quickly, it was really about life in Universe 25. Indeed, the article ends with a full-page photo of mice aggregating around Calhoun's feet in Universe 25 with a legend that reads, "Author plays Pied Piper to his overcrowded and hence socially deprived mice. They have never learned normal adult ways, so they follow his feet with the thoughtless curiosity of babies."

Before the mice enter into the article, Calhoun reminds readers of the dramatic, sometimes suicidal, "march to the sea" undertaken by lemmings (*Lemmus lemmus*) in some Norwegian populations. It begins in the mountains: "Long before the lemmings begin to overtax their food supplies a general state of unrest develops. More and more of the animals forsake their normal sedentary way of life, wandering about aimlessly, upsetting still sedentary associates," Calhoun writes. "When too many young lemmings . . . survive to adulthood, increased contacts among them lead to anxiety, strife and an unsettled way of life." As one group of lemmings moves down the mountain, it unsettles another, eventually leading swarms of lemmings to the valleys and finally to the sea, where many drown.

The lemmings' march to the sea took Calhoun back to the late 1940s, when he studied rodent communities on Mount Desert Island, near JAX. "As each strip of [his] animals moved toward the void [that Calhoun had created at Mount Desert], their neighbors occupied their vacated homeland, moving in as if to escape from social pressures of others still further out." But that was decades in the past. "Perhaps you have heard of my 'horrible mousery,' my 'paradise' for house mice," Calhoun asks *Smithsonian* readers, referencing Sartwell's article a year earlier. "It really is a mouse utopia: it provides for all their physical needs. But unlike the lemmings' ranges of Norway, the walls around this utopia allowed no escape when the mouse population began to exceed tolerable limits."

The next two pages of the article summarize what was happening in Universe 25, including the Beautiful Ones, described as "hollow shells, not participating in any social life with the favored few." And Universe 25 shed light on what lemming life might be like if they couldn't emigrate: "In [a] last frenzy of reproduction," Calhoun writes, ". . . mice were spewed out into this closed environment. A replica of 'spaceship Earth'?" If so, the prospects were not good, for the population of mice in Universe 25 "seems doomed," according to Calhoun. "They have already committed suicide." We don't have to follow that path, Calhoun next tells his readers, but we might. As a way to avoid self-destruction, Calhoun was to tap into and amplify our creativity to tackle overpopulation. Ideas generated in conceptual space are our only hope. "Periodically [man] has been faced with crises, not of numbers, but of ideas. Old ways became outmoded, new ones emerged. We have always had the choice which would survive, which would be allowed to 'commit suicide,'" Calhoun writes in an evangelical tone. "Sometimes the new ways drew the lot of suicide and areas of the world were left to stagnate and die. Sometimes areas adopted new ways that enlarged human potential." The choice was ours: "Whenever we fail to produce ideas and utilize them," Calhoun writes toward the close of the *Smithsonian* article, "we commit suicide."[9]

13

DEATH SQUARED

John Calhoun felt lucky to call NIMH his research home. When he was offered a job as chairman of a new Department of Ecology and Behavioral Biology at the University of Minnesota, he politely declined, knowing that with the dizzying array of disciplines that he was tapping into to try to solve the population crisis, few, if any, other institutions would give him the freedom he had at NIMH. "The National Institute of Mental Health has mothered and encouraged me," he wrote in "The Population Crisis Leading to the Compassionate Revolution and Environmental Design," a paper he published in an avant-garde publication called the *World Journal of Psychosynthesis*. "I have rejected, though not lost sight of, my 'father,' the discipline of animal ecology. When I now speak of the NIMH, it is not solely of it in the institutional sense, but more as its embodiment of the role behavioral sciences, including psychiatry, may play in an emerging future."

Slipping into the poetic mode that he thought all scientists should, on occasion, slip into, Calhoun went as far as speaking of "KANIM," where "KA" was "spirit of; essence of; genius of; grace; elan vital" and "KANIM" was "the spirt and purpose of NIMH." Calhoun's dream was for "those who share the ecological perspective with me [to] inseminate the behavioral sciences to produce a KANIM capable of initiating the Compassionate Revolution." *If* that compassionate revolution were to come about, Calhoun predicted it would encompass a "world-wide communication network in which individuals serve as nodes . . . 3.5 billion adults." When that happened, it would "mark the termination of the past 50,000-year epic of evolution."

KANIM, as well as its nonspiritual counterpart, NIMH, allowed for travel. In early May of 1972, Calhoun and his wife, Edith, headed to the Sea Palms Resort Hotel on Saint Simons Island, Georgia. NASA had selected that locale to host a Forum for Speculative Technology, and in a sign of just how much his work was in public eye, Calhoun was one of a select group of sixty in attendance. The theme of the forum was "Information: Interactions in the Next Generation." While it is not entirely clear what expertise NASA saw Calhoun bringing to the table, the fact that the flyer for the forum announced that "aeronautical and space systems will most certainly be enhanced by future improvements in information handling, retrieval, and storage systems . . . and communication systems" suggests it was his work on invisible colleges, appreciative systems, and electronic prostheses that led to the invitation. What we do know is that Calhoun believed he was brought in as an "idea generator" employed "to explore the social, cultural, political and biomedical implications of current and anticipated breakthroughs in communication systems and technologies."

It was an eclectic lot that gathered at the Sea Palms Resort. James C. Fletcher, former head of NASA, was there, as were Edgar Mitchell, lunar module pilot of Apollo 14, and physicist Wernher von Braun, director of NASA's Marshall Space Flight Center. They were joined by many others, including Stuart Umpleby, a former student of von Foerster of doomsday model fame, and Marvin Minsky, an MIT computer scientist and early founder of the field of artificial intelligence, who was on a panel charged with examining "What kinds of intelligence might be built into non-human systems?"

It wasn't just scientists that NASA brought to Saint Simons Island: Frederik Pohl, author of the 1953 science fiction book *The Space Merchants*, which tells the story of an advertising agency's plan to colonize

Venus to market products in space, was also part of the forum. Pohl, like Calhoun, was an attendee rather than a presenter, meaning they took part in the discussions that followed each of the six, multilecture sessions. But science fiction writer Arthur C. Clarke, whose 1951 story "The Sentinel" was the basis for Stanley Kubrick's 1968 classic film, *2001: A Space Odyssey*, and who was a commentor called in by Walter Cronkite when Apollo 11 landed on the moon, was not only an attendee but also a moderator for a forum session titled "Definition of Socio-Technical Information Problems." If NASA was attempting to think outside of the box by bringing in people like Pohl, Clarke, and Calhoun, it worked. Calhoun's daughter Catherine recalls her mother telling her that at some point during the forum she walked into a room and saw her husband and Arthur C. Clarke standing on their heads. Edith didn't mention what the two were talking about, but it would not be surprising if it had something to do with communicating in zero gravity.[1]

The following month Calhoun was on the road again, this time to London. The Royal Society of Medicine had asked him to be a speaker in a one-day (June 22) symposium they were organizing on "Man in His Place," and Calhoun viewed this as a chance to give a comprehensive overview of Universe 25, along with its implications for both mice and men. The other speakers at "Man in His Place" were a diverse lot that included mathematician Jacob Bronowski, whose popular thirteen-part BBC television series, *The Ascent of Man*, would begin broadcasting the following spring; James Weiner from the London School of Hygiene and Tropical Medicine; Sir Fraser Darling, an ecologist who had done a long-term study of red deer on the Isle of Rum; and Fred Sai from the International Planned Parenthood Federation.

"Our aim today is to see what we can learn by looking at some of the similarities between the problems that are faced by human and animal populations," J. Z. Young, one of the symposium organizers, told the audience. "There will be speakers who are very well able to tell us about those problems from either one or both sides and we shall try to see how far they can be met by similar solutions. . . . Let us rejoice in the fact that we live at a time when ever-fresh sources of information abound, and when people are beginning to realize the need for closer and closer co-operation for survival."

Calhoun was the first lecturer in the symposium, and Young's introductory remarks were no way to prepare the audience for what was to follow. Fellow symposium attendants, as well as members of the Royal Society and reporters from the *Daily Telegraph* and *Daily Mirror*, might have sensed from the title of the talk—"Death Squared: The Explosive Growth and Demise of a Mouse Population"—that it would not be a run-of-the-mill lecture, but no one could have envisioned just how unusual an experience they were in for.

"I shall largely speak of mice, but my thoughts are on man, on healing, on life and its evolution," Calhoun began, which on the surface is an unusual opening line for a Royal Society lecture but, given that the symposium was on "Man in His Place," not completely out of place. What Calhoun next said must surely have struck the audience as unusual, not to mention uninterpretable: "Threatening life and evolution are the two deaths, death of the spirit and death of the body," he proclaimed. "Evolution, in terms of ancient wisdom, is the acquisition of access to the tree of life." One can only imagine what that august audience thought when Calhoun continued: "This takes us back to the white first horse of the Apocalypse which with its rider set out to conquer the forces that threaten the spirit with death. Further in Revelation (ii.7) we note: 'To him who conquers I will grant to eat the tree of life, which is in the paradise of God.'"

If listeners were expecting an explanation, what they got instead was more scripture. "This takes us to the fourth horse of the Apocalypse (Rev. vi.7)," Calhoun said. "I saw . . . a pale horse, and its rider's name was Death, and Hades followed him; and they were given power over a fourth of the earth, to kill with the *sword* and with *famine* and with *pestilence* and *by wild beasts* of the earth. This second death has gradually become the predominant concern of modern medicine." Then, finally, some sense of what on earth was going on with this odd

American: "Let us first consider the second death," Calhoun said. "The four mortality factors listed in Revelation have direct counterparts (with a division of one of them to form a total of five) in the ecology of animals in nature. I shall briefly treat each of these five mortality factors, and then discuss the steps taken to eliminate, or drastically reduce, the impact of each." At that point Calhoun clicked the projector and the slide on the screen read:

As in Revelation vi.8	*Ecological expression*
(1) Sword	(1) Emigration
(2) Famine	(2*a*) Resource shortage
	(2*b*) Inclement weather (and fire and cataclysms of nature)
(3) Pestilence	(3) Disease
(4) Wild beasts	(4) Predation

Emigration, *resource shortage*, *inclement weather*, *disease*, *predation*: at last, some terms that scientists in the audience would have understood. What Calhoun went on to say was that Universe 25, by design, had none of the ecological versions of the scourges in Revelation, and at its inception, it was nothing less than a "utopian environment constructed for mice." It didn't stay such for long, Calhoun explained, as he walked the audience through the different phases the population of mice went through: phases that those in the audience who had read Calhoun's Casey Barn *Scientific American* article ten years earlier would have recognized.

Universe 25, Calhoun explained, was born on July 9, 1968, when four males and four females were placed into his mouse Shangri-la. During phase A, which lasted three and half months, no pups were born and there was "considerable social turmoil among these 8 mice until they became adjusted to each other and to their expanded surroundings." In phase B, the population experienced exponential growth, in which it doubled about every fifty days and skyrocketed to more than six hundred individuals. During phase B there were fourteen established groups, each with about ten mice, including a dominant male and multiple females and subordinate males. In total, these groups produced 470 pups, which experienced what Calhoun described as "good maternal care and early socialization." Lest the audience members began to think that perhaps paradise was on the rebound, Calhoun explained there was a problem for these well-adjusted young ones, as their "number [was] far greater than would have existed had the normal ecological mortality factors functioned."

Day 345 marked the start of phase C, which lasted for six months. A behavioral sink was in place now, with mice clustering at certain feeders and avoiding others. The growth rate plummeted in this phase, with the time to double population size shifting from fifty-five days to almost five months. Calhoun explained to the audience what had happened: in nature, there was rarely, if ever, three times as many healthy young waiting to fill the social position of the adults, as was the case at the end of phase B: predation, disease, lack of food, and the vagaries of living in the wild saw to that. And any excess young that might survive in such an overcrowded natural population would have emigrated, which they simply could not do in Universe 25. This led to many young males in Universe 25 fighting for dominance, and "[those] who failed withdrew physically and psychologically; they became very inactive and aggregated in large pools near the centre of the floor of the universe."

As younger females matured, they also faced problems. Many were forced to higher, less-preferred apartments in each of the sixteen cells. Pregnancy rates dropped, and many females who became pregnant never gave birth. Females who did give birth were poor mothers, who "transported their young to several sites, during which process some were abandoned." What all this meant was that pups born during phase C "started independent life without having developed adequate affective bonds," and that the "maturation of the more complex social behaviours such as those involved in courtship, maternality and aggression failed." When these pups matured, Calhoun explained, they were the "beautiful ones . . . capable only of the most simple behaviours compatible with physiological survival. . . . Their spirit has died ('the first death')." In case he had not conveyed his message in stark enough terms, which seems hard to believe, Calhoun pronounced that "for all practical purposes there had been a death of societal organization by the end of Phase C."

Phase D commenced on day 560. Very few mice were born between day 560 and day 600, and then, starting on day 600, not a single pup born survived. By day 920 no females were getting pregnant. On the day Calhoun spoke to the Royal Society, the population of mice in Universe 25, which had peaked at 2,200, now sat at 100 females and 22 males: "Projection of the prior few months of exponential decline in numbers," Calhoun noted, "indicates that the last surviving male will be dead on May 23, 1973, 1780 days after colonization.

Calhoun in Universe 25 (photo appeared in the paper that emerged from his "Death Squared" talk to the Royal Society of Medicine). Many of the Beautiful Ones are on the floor, opposite Calhoun. A behavioral sink was in place with mice clustering at certain feeders (*far left and far right*) and avoiding others. Photo credit: Yoichi R. Okamoto.

The population will be, reproductively, definitely dead at that time." Though both the audience and Calhoun didn't know it then, Calhoun's prediction was close to the mark. The proceedings of the symposium were published in 1973, and in his paper, Calhoun noted that "at final editing of this paper on November 13, 1972 (Day 1588) the inexorable decline brought the population to 27 (23 females and 4 males, the youngest of which exceeded 987 days of age)." Mice never reproduce at so old an age. Universe 25 was dead.

The question-and-answer session that followed Calhoun's talk began with Professor Young, the chairman of the session, asking Calhoun whether he thought pollution—remains of dead mice, feces, and urine—played a role in his findings. Calhoun said he thought not: at least not *that* type of pollution. Based on "about a million observations," he replied, it was clear to him that "pollution was social, in that there were too many interacting elements, exceeding the social system's capacity for incorporation of new individuals." Then, in reply to a question regarding overpopulation in humans, Calhoun began discussing his ideas on the implications of Universe 25 for humans, noting that "despite the thousandfold increase in human numbers since the beginning of culture, some forty to fifty thousand years ago, there had been no change in effective density. The reason for this . . . was that man had discovered a new kind of space, conceptual space, which enabled man to utilize ideas in order to mine resources and guide social relations." But there was a problem: "there was a breaking point for this process, at which time physical density might overwhelm man's ability to utilize conceptual space in order to cope with increasing numbers and it was that breaking point which we might be rapidly approaching." The chairman then "interrupted Dr Calhoun and said that he ought to be careful about this. He had been discussing mice," not humans. Calhoun was not about to back down, he "thought this to be a major trap." He believed "mankind was on a knife-edge which could shift in a couple of directions. One behavioural shift on the human side could be comparable to 'beautiful one' mice: individuals capable of the routine of life, but with loss of creativity and an inability to live under challenge."

The Q and A session was short, and there was only time for a few queries. No one asked Calhoun whether he was concerned about the particular strain of mice—the Balb/C inbred strain—he was using in his universes, but Frank Sartwell had when he had written "The Small Satanic Worlds of John B. Calhoun" for *Smithsonian* two years earlier. One "standard argument," Sartwell told Calhoun was "the experimental mice are not only not human beings, they really aren't very good mice. They are so thoroughly inbred that they simply don't react like any other animals." Calhoun replied that as far as he knew "wild species suffer the same reactions as [his mice], sometimes even in exaggerated form." What's more, using NIMH's Balb/C inbred strain, Calhoun was assured of healthy mice, as NIMH had many safeguards in place to minimize the spread of disease. Still, Balb/C mice were only one of dozens of different strains of mice used in laboratory work at the time. At the 1966 winter meeting of the Society of American Zoologists, Charles Southwick and Linda Clark presented a talk in which they discussed work they had done comparing aggressive behavior and exploratory behavior across fourteen different strains of mice, including the Balb/C strain. Calhoun wasn't at the meeting, and it's not known whether he saw the one-paragraph abstract of the talk that appeared in the *American Zoologist*, but what Southwick and Clark found was that, compared to the other thirteen strains, "Balb/C's were aggressive but relatively inactive in exploratory . . . behavior." Given that exploration played a key role during phase A of population growth in Universe 25, and that aggression was important in all phases of growth, some caution should be taken when generalizing the results uncovered in Universe 25. That said, the fact the dynamics of population growth in Universe 25 were remarkably like those in the rats of the Casey Barn fame can't be ignored.

The day after the "Man in His Place" symposium, the *Daily Mirror* ran a story entitled "1984 May Be Year of Doom," and the same day, the *Daily Telegraph* printed not one but two stories: "Mice Point Way to Doom in 1984, Says Scientist" and "Sad Elderly Mice." When the last mouse in Universe 25 died in early January 1973, the *Washington Post* ran a long story titled "Ten Boxes of Dead Mice Could Be Us: Is Modern Mankind Becoming a Giant Colony of Mice?" which began "the newspaper death notice could have read: MOUSE, THE LAST, on January 8, 1973 at the laboratory of Brain and Behavior Evolution; of terminal overcrowding; no survivors; no services; no matter." The article, which was on the front page of the B section of the paper, even had a photo of ten boxes of dead mice from Universe 25. Calhoun told the reporter, Thomas Huth, who three years earlier wrote the *Washington Post* story that stirred Senator Packwood to discuss Calhoun's work on overpopulation with his colleagues in United States Senate, that there was no mourning when the last denizen of Universe 25 left the world of the living: "You can't identify with nothing. . . . These animals were nothing. They weren't mice. . . . They had motion but it had no meaning. It's the minimum motion needed to keep the machine going."

Everyone seemed interested in Universe 25: Calhoun received hundreds of letters, and on the afternoon of March 19, 1973, he found a "memorandum of call" note on his desk that read "Mr. Paul Block of *The Tonight Show* (Johnny Carson)" had called. Many years later Calhoun annotated the memorandum with "this never materialized." *The Tonight Show* appearance may not have materialized, but many other things did. *CBS Morning News* host Nelson Benton took a camera crew and spent time visiting Universe 25 and interviewing Calhoun. The piece, which ran nearly five minutes long—an eternity for a sixty-minute national news show—aired on March 1, 1973, on both the morning news and the evening news.

When Calhoun's symposium paper, in which he was mislabeled as an MD in the byline, was published in the hallowed pages of the *Royal Society of Medicine* in 1973, the *Boston Evening Globe* ran a front-page story called "Of Mice and Men and Paradise Lost," the *Los Angeles Times* countered with "Easy Living in a High Rise, but Paradise was Brief," and the *Washington Star-News* published "Of Mice and the Welfare State." The authors of a major introductory biology textbook were in the process of revising the book for the next edition and wrote to Calhoun requesting photos from "your Universe 25 or 'dead city' . . . illustrating some of the social interactions that took place in your experiments." Even *Good Housekeeping* covered Universe 25, albeit briefly, in their "Results of Overcrowding." Adjacent to an advertisement for a hair dye called Happiness, the editors wrote that "where overcrowding leads is one of the most complex and important scientific problems we face—one intriguing experiment done with mice suggests that it can lead to community suicide." After giving a brief overview of Universe 25, the *Good Housekeeping* story ends with a caveat: "Human beings are not mice and a play experiment with children offers an interesting contrast," at which point they describe work done at the Ohio State College of Medicine in which four- and five-year-old children adjusted to overcrowded conditions by being especially gentle and friendly to one another.[2]

Not long after he presented "Death Squared" to the Royal Society, Calhoun published his second article in *Smithsonian* magazine: a disturbing piece, titled "Plight of the Ik and Kaiadilt Is Seen as a Chilling Possible End for Man." The article was largely a commentary on Colin Turnbull's 1972 book, *The Mountain People*, about the Ik people of Uganda. A year after Calhoun wrote this article, a strong critique questioning the veracity of some of Turnbull's claims was published, and many similar critiques have been made over the years. But at the time, Calhoun took Turnbull at his word and worked from there.

In the *Smithsonian* article, Calhoun, following Turnbull, described how Ik society collapsed from a peaceful hunter-gatherer society after the Ugandan government banned the Ik from hunting in their native Kidepo Valley and moved them to a nearby mountain, where they were forced to rely on farming very poor land. Once Calhoun introduced the Ik, or at least Turnbull's rendition of them, to the readers of *Smithsonian*, he made some rather unsettling comparisons between the Ik people and the mice in his Universe 25. As he described the collapse of Ik society, Calhoun wrote, "Strike No. 1: The shift from a mobile hunter-gatherer way of life to a sedentary farming way of life made irrelevant the Ik's entire repertoire of beliefs, habits, and traditions. Their guidelines for life were inappropriate to farming." So too for the inhabitants of Universe 25: "Strike No. 1 against these mice," Calhoun wrote, was that "they lost the opportunity to express the capacities developed by older mice born during the rapid population growth. After a while they became so rejected that they were treated as so many sticks and stones by their still relatively well-adjusted elders."

Strike 2 against the Ik: "They were suddenly crowded together at a density, intimacy and frequency of contact far greater than they had ever before been required to experience." So too for the mice in Universe 25, Calhoun explained, introducing readers to the idea of the behavioral sink and the Beautiful Ones, in whom "all true 'mousity' was lost." If only he had known of the Ik when he spoke at the Royal Society symposium, Calhoun told the readers of *Smithsonian*, he could have summoned them to his defense: "The chairman admonished me to stick to my mice; the insights I had presented could have no implication for man. Wonderful if the chairman could be correct—but now I have read about *The Mountain People*, and I have a hollow feeling that perhaps we, too, are close to losing our 'mountain.'"

The same year that he presented "Death Squared" to the Royal Society and wrote of the Ik and Universe 25 in *Smithsonian*, Calhoun was nominated for a Nobel Peace Prize. Years after her father's death, Catherine Calhoun discovered an index card in his files that read, "Dick Wakefield in 1972 sent in a letter of nomination for me for Peace Prize. When I visited Stockholm in 1972 . . . the Peace Research Institute was aware of this." When Catherine contacted the Nobel Peace Prize Forum, they confirmed that "one professor nominated your father one year." While the Forum would not release the name of the nominator, there is no reason to doubt her father's attribution: Wakefield, an ecologically oriented urban planner, was a close friend of Calhoun's fellow Space Cadet Ian McHarg and was also acting director of NIMH's Center for Studies of Metropolitan and Regional Mental Health Problems.[3]

In the winter of 1972, Eleonora Masini, a sociologist and futurist at the Applied Research Institute (Istituto Ricerche Applicate) in Rome invited Calhoun to be a speaker at the Special World Conference on Futures Research to be held in Rome September 25–30, 1973. Masini felt it "extremely important" that Calhoun "explain the possibility of a prosthesis to the brain to the participants." Calhoun was honored at the invitation and accepted straightaway. He suggested that legendary Italian movie director Roberto Rossellini, who had once invited him to China and India for a film Rossellini planned on population growth, be invited, which he was: the following year Rossellini produced *The World Population: A Question of People*, a two-hour documentary for UNESCO.

"Take One Step," a special musical piece composed by Roderic Gorney, with lyrics that included "We will take one step closer to survival," welcomed the participants to the Rome meeting, as did Pope Paul VI. Using the ecclesiastical version of the royal *we*, the Pope told those gathered, "We are aware of the general theme of the Conference: the study of man and his future. . . . As scholars . . . you are in a position to offer the coming generations authoritative perspectives of development, and to contribute to the improvement of human life." After reminding everyone present that "in this field the Church, as the bearer of a transcendent and revealed doctrine, certainly has something to say," the Pope sent them off with his blessing: "We wish you well in your work, which is both forward-looking and courageous. You are pioneers, blazing a trail for future generations. We do not doubt that your commitment will be fruitful and appreciated." At some point after his welcoming remarks, the Pope did a meet and greet that included shaking hands with Calhoun.

Calhoun, listed as "an expert in behavioral systems," was a member of Conference Group 5, which was charged with addressing how to facilitate "The Participation of All People in Physical and Ecological

Survival." Group 5 included statisticians, biologists, industrial engineers, philosophers, and philanthropists, and one recommendation the group made to the larger conference was to find "sources with an interest in global survival and an improved human future to create an 'innovation bank' [to provide] seed money, grants, loans . . . for innovative experiments."

When Calhoun's time came to speak, he presented a talk with this unwieldy title (using "℞" in place of *R*): "Metascientific Research: A Proposal for a Demonstration Effort to Evaluate 'Research on Synergistic Research' as an Organizational Methodology for Promoting Scientific ℞evolutions in the Understanding of the Complex Dilemmas Characterizing Biomedical-Behavioral-Environmental Phenomena in Contemporary Society." Masini had asked him to discuss electronic prostheses and their role in the future, and Calhoun did that and more, opening his talk with a stark reminder: "The earth is now involved in an incipient evolutionary phase shift unmatched in magnitude by any prior ones. Its accomplishment will involve major changes in values, purpose . . . all in the direction of increasing the synergistic interrelations among ideas, people, and organizations." The talk was even more unwieldy than its title, but at its core Calhoun preached that electronic prostheses and an understanding of appreciative systems, as well as invisible colleges—hotbeds of what he called "metascience perspective(s)"—would all play critical roles in that evolutionary phase shift.

Even before the conference was over, Calhoun wrote the organizers to suggest what needed to happen next. Looking out the window from his hotel room, Calhoun penned "A Villa Tuscolana Statement of Intent to Initiate 'the Four Catalytic Years' 1974–1977." There he mused: "From the heights of this villa Cicero gained visions appropriate to his times. Today we seek visions that spread outward from our time. . . . The world is on the verge of the most profound transformation in all the history—in all of evolution." Calhoun had come to think that if we didn't somehow make significant strides toward solving the population problem by 1986, we were unlikely to solve it at all.

"If the 13 years from 1973 to 1986 are the crucial ones," he wrote, "many processes affecting awareness, participation and change must be set in motion. The present proposal of intent is to develop the four years of 1974, 75, 76 and 77 as 'the four catalytic years.'" That change would involve the development of a "visionary declaration of ℞evolution [that] . . . will speak *from* all people . . . [and] catalyze those actions needed to be completed by 1986." That declaration would in part be drawn by invisible colleges, which would "increase the effective contribution of *all* people to the decision processes affecting the manner in which current changes in their ways of living determine the long range future." But it wasn't only invisible colleges that would be key, as Calhoun also suggested that the Smithsonian Institution act as a central hub for action and that "its possible role will be explored."[4]

More and more Calhoun wrote of ℞evolution. In 1973 he published a paper, "℞evolution, Tribalism, and the Cheshire Cat: Three Paths from Now," with the opening sentence, "Examination of the time course of the change in numbers of experimental populations of mice, and of the world population of man, reveals abrupt and apparently irreversible phase changes in the processes governing evolution." Many pages were devoted to the mice in Universe 25, but, as for man, there were only the three choices listed in the title of the paper. One, the Cheshire Cat path, led to extinction: along that path the human equivalent of the Beautiful Ones—"aware of less and less as numbers of individuals exceed the upper optimum"—would become more common.

"Beyond this nonchoice," Calhoun wrote, were the paths of tribalism and ℞evolution. Tribalism, in his view, might lead to zero population growth, but was "accompanied by a pervading traditionalism that establishes invariant patterns of social, technological and ideational function." This, too, was a nonchoice for Calhoun. Invariance was not in his nature: elsewhere he wrote, "I for one would not like to be a lung fish in Lake Manyara in Tanzania. For millions of years they have maintained a constant way of life avoiding crises. . . . They have learned to hide and not to cope. As the lakes dry up each lung fish burrows down into the mud, forms a mudball cocoon . . . sleeps out the trying, drying period. We too could devise our solitary cells of intellectual isolation to wait out each and every crisis."

Calhoun believed the only viable solution was ℞evolution, wherein we "choose to design further evolution . . . by an increase in the effectiveness of technological prostheses for information processing beyond that possible by biological and social brains." Choosing that path, Calhoun knew, would face resistance from the status quo. Twenty years after writing this paper, Calhoun reread it and marked it with annotations along the margins. To the right of the following sentences, he jotted down, presumably for posterity, the words "JBC's core philosophy": "Our success in being human has so far been derived from our honoring deviance more than tradition. . . . Now, we must seek diligently for those creative deviants from whom alone will come the conceptualizations of an evolutionary designing process which can assure an open-ended future—toward whose realization we can all participate." Calhoun was well read, but it's not clear whether he knew that John Stuart Mill made a similar argument more than two hundred years earlier: "The amount of eccentricity in a society has generally been proportional to the amount of genius, mental vigour, and moral courage which it contained," Mill wrote in *On Liberty*. "That so few now dare to be eccentric, marks the chief danger of the time."

In defusing the population bomb, Calhoun was convinced that time was of the essence, and so everyone, from politicians to high school students, needed to be on board. Senator Packwood had invoked Calhoun's work for the cause but never directly contacted him. In March of 1973, Congressman Barry Goldwater Jr. did. A Republican from California, and a member of the House of Representatives Committee on Science and Aeronautics, Goldwater wrote Calhoun that he had read of his "1968–1973 experiment with rat populations and overcrowding," and that "a number of the 'effects,' as stated in the article, appear to coincide with certain theories now being evaluated by land-use planning consultants working with [Goldwater's] staff." Goldwater asked Calhoun for copies of his work and requested "the opportunity to have my consultants confer with you concerning the possible relationships between your findings and our investigations." Calhoun wrote back a week later, and after diplomatically correcting Goldwater that the studies he read about were on the mice, not rats, he noted that he was enclosing a dozen reprints for the congressman and his staff to read. Calhoun reminded Goldwater: "Although our research has been solely with animals we have sought to gain insight into the possible implications of our findings for the human scene."[5]

As for high schoolers, in 1973, two years after *Mrs. Frisby and the Rats of NIMH* was on bookshelves everywhere, young readers again met the rats (and mice) of NIMH, this time in nonfiction form. In that year Calhoun gave a teacher in Boulder, Colorado, permission to "adapt 'Death Squared' in the development of a high school curriculum," and A. H. Drummond, a high school teacher at St. Paul's School in Garden City, New York, published his book *The Population Puzzle: Overcrowding Among Animals and Men*. Aimed at early high school students, the book opens with a dramatic photo of rush hour on the Tokyo subways, with men in uniforms and arms extended forward standing before open train doors: "Because there are too many people train passengers are packed together by 'pushers' during rush hours in Tokyo. All aboard!"

Drummond painted a bleak Malthusian picture of times to come for his young readers. To help them understand what "man may be doing to himself by letting overpopulation take place"—what Drummond called the population puzzle—he asked for their patience: "Let's go back and look at some of the experiments that have provided clues to the solution of this puzzle." No experiments detailed in the pages that followed played a bigger role than those of Calhoun. Drummond not only read Calhoun's papers, but he had also written to Calhoun at NIMH and asked for any information that he could provide him for his readers. Calhoun was thrilled at the idea that fledgling thinkers would be exposed to his work and sent Drummond "a wealth of material related to his work."

Drummond took his readers to the Towson enclosure and showed them, in a full-page photo, how "male rats harassed a female until she retreated into her burrow." Next, he transported them to the Casey Barn and told them of "'probers,' 'pansexuals' and 'drop-outs (zombies)'" and Calhoun's ideas on optimal group sizes of twelve and behavioral sinks, which he oversimplified to "a place where undesirable behavior takes place." Later in the book, Drummond takes his teenage readers to Universe 25. Along the way, he challenges students to contemplate whether this will "happen to overcrowded human populations." Assuming that "after reading about Calhoun's experiments, you may want to agree," Drummond cautioned them that "it is dangerous to leap to conclusions when animals and humans are compared," even though this is exactly what he had done all throughout *The Population Puzzle*. Drummond wasn't through with Calhoun yet. He next introduced von Foerster's doomsday model and Calhoun the "℞evolution[ist]," including Calhoun's ideas on human evolution and how we had "evolved the ability to use conceptual space . . . the world of ideas" that ultimately might, but then again might not, save us from the population explosion.

"You decide," Drummond told his readers. "Is [Calhoun] a prophet of doom or a prophet of hope?" A heady decision for anyone, let alone a high schooler, to make.[6]

14

I PROPOSE TO MAKE AN APE OUT OF A RAT

In early 1974, Calhoun sat down with a stack of index cards, and on each one he wrote down a journal in which his work had been cited. It was an impressive, diverse lot that included journals such as *Evolution*, the *American Journal of Sociology*, *Animal Behaviour*, *Ecology*, the *American Psychologist*, the *American Anthropologist*, the *American Biology Teacher*, the *Archives of Medicine*, *Futures*, the *Journal of Personality and Psychology*, *Hormones and Behavior*, the *Journal of Theoretical Biology*, and *Current Science*. Along with widespread citations to his work came an invitation to speak at a conference on "Exploding Cities," cosponsored by the United Nations and the *Sunday London Times*. The conference, part of the launch of the United Nations World Population Year, was held in London during the first week of April 1974. The conference organizers, at the suggestion of Calhoun's friend Roberto Rossellini and Dr. Jonas Salk of polio fame, asked Calhoun to be part of a discussion group on human values and the urban environment.

"I want to talk to you today on a theme," Calhoun began his remarks at the London conference, "which I shall call 'the universal city of ideas.'" To set the mood, Calhoun told the audience that the day before the conference he had stopped at a gift shop on High Street, where he saw three statues: one of a dodo, one of a great auk, and one of a passenger pigeon, all species that are extinct. "Accompanying each there was a message," Calhoun said. "Are we next?" Perhaps, he answered. Creativity was the only way forward. In the evolutionary past, environments were constantly changing, and "as environments changed, life changed—but not always," Calhoun explained, as there were some species who adopted the rule "Just find a niche where the old way of life can be maintained. Build a wall that wards off all threats." It wasn't just the mud-burrowing lung fish in Tanzania he liked to trot out as an example on such occasions: "We, too, wrap ourselves in warm blankets of tradition," Calhoun said, "to sleep in a capsule free of the volatile crisis pounding outside." But, he made clear, we are by no means obliged to remain coddled in our own swathe cloth. The way out was fostering new ideas. The strong among us, he argued, "remain where conditions are most salubrious to preserving the old lifestyle." But others were more likely to emigrate "bodily, behaviorally or intellectually," Calhoun told his audience. As a result, he said, "Their brains grow. They gain the capacity to acquire, to store, to transfer and to transform information." We need to foster those intellectual migrants, Calhoun proposed, because they're the path to avoiding extinction. And we need ways of connecting the ideas generated by such individuals into a "world brain in which each of us is a neuron or node." Of course, this was the view of one man, Calhoun told those gathered, but "such is the vision from the Maryland countryside."[1]

Buoyed by the recognition he was receiving and by both Senator Packwood's and Congressman Goldwater's interest in his work and its implications for humans, Calhoun penned a letter to President Nixon. Sidestepping the Watergate hearings that were underway at the time, he opened the letter by telling the president that at Olmsted Restaurant in Washington he had bumped into Robert Finch, a former secretary of the Department of Health, Education, and Welfare and counselor to the president. He had invited Finch to the NIH field station, Calhoun told Nixon, "to come out into the quiet of the countryside and engage in reflective dialogue with me whenever the pressures of office became burdensome," but Finch declined. Perhaps the president might come, as "site and solitude and circumstance are the midwife of visions," Calhoun wrote Nixon, and vision was what was needed to save humanity from the population explosion.

The population crisis, Calhoun warned the president, was the gravest threat humanity had experienced in fifty thousand years. "How we resolve this crisis will determine the course of history for the next 100 thousand years. . . . Your decisions, by default or intent, will make the difference." Next came an invitation to spend a week at what Calhoun portrays as an Elysian-like NIH field station, where "the starkness of winter will unfold through spring flowers to verdant summer foliage encouraging the unfolding of our thought." And, for good measure, Calhoun added that the president need not worry about his mode of transport, for out at the NIH field station there was more than "adequate helicopter landing space."

Whether Nixon ever saw Calhoun's letter is unclear, as the only response issued from the White House was a pro forma notecard from Roland Elliot, special assistant to the president, that read, "The President appreciated your thoughtful message." Generating interest in congressmen and senators was one thing, Calhoun now understood, but garnering interest from the White House, as calls for impeachment were growing louder and louder, was another.[2]

Calhoun found he had bigger problems than being ignored by the president. The budgets for the projects he was working on were being cut, as part of a general decrease in spending on behavioral studies, particularly behavioral studies on nonhumans, at NIMH. "I began to retrench as an accommodation," he wrote, and by the end of April 1974, "I had resigned myself to spend the rest of my career working alone, tidying up loose ends of ideas developed." Things only got worse, and by the next month, Calhoun described himself as in a state of "deep depression." It had been twelve years since he had taken his sabbatical to the Center for Advanced Study in the Behavioral Sciences (CASBS) in Palo Alto, and he came to think another stretch might be just what he needed to raise his spirits and allow him to move forward.

In May of 1974, Calhoun applied to go on a sabbatical of sorts from June 1 to December 31. In that application he noted that his time at CASBS during his first sabbatical had provided a "period of reflective thinking, divorced from the pressures and diversions" that permitted him "to develop formulations capable of encompassing or interrelating the many complex phenomena that impinge on [his] research into population and environment related problems." Six months of reflection now, he proposed, would allow him to do something very similar. He would use the time, he told the higher-ups at NIH, to work through all the rat and mouse data he hadn't had a chance to analyze, and to map out a new series of experiments, taking into account the budget cuts that were in place. He also planned—though he didn't mention it in the sabbatical request—on spending time thinking about invisible colleges, as well as how to "initiate a frontal attack on development of the prosthetic brain," and he would give more serious thought to writing a book on what the rats and mice had to say about population dynamics—theirs and ours.

Calhoun's plan would require "a total immersion . . . for long periods of time without distraction," and so he told NIH that he would remain put, taking the sabbatical at home rather than spending time away in a CASBS-like setting. He proposed spending Tuesday to Friday in his home office, during which time "no phone calls [would] be accepted." On Mondays he would go to the field station to work on administrative functions for the new studies he would design during the sabbatical. As for the day-to-day care of the rats and mice while he was on sabbatical, Calhoun made it clear that the others on his research team were more than capable of taking care of that, so that the rats and mice would be ready when he was. A week later, the sabbatical was approved: "There is no question," read the confirmation letter, "but that his . . . research program requires this kind of concentrated study and analysis, free of interruptions."[3]

Calhoun began his sabbatical one month after his fifty-seventh birthday. He had no intention of retiring any time soon, but he knew he was approaching the latter part of his career, and so he used part of his sabbatical to write a thirty-page document titled "Last Cycle of Research Inquiry at NIMH Guided by John B. Calhoun, 1975–1982: A Proposal." That proposal opened with Calhoun discussing his metascientific approach to the study of population growth and dynamics. "Metascience builds on the strengths of basic normal science," he wrote to the higher-ups at NIMH who would read this document. "It differs from basic normal science in its strategy of employment of a large number of variables, whose interaction can never be completely predicted at the onset of a planned effort." What this approach provided was the power to dig deep into complex systems and to be better prepared to handle "processes and phenomena [that] could never be anticipated from considering results of less complex studies." Though others were skeptical—one colleague told Calhoun, "There are so many variables in your research that you can't possibly draw any conclusions about anything"—the metascientific approach, Calhoun believed, was critical for understanding population growth and decline both in animals and humans. If administrators needed proof of that, they need only look at Calhoun's studies of mice and rats, and the implications that emerged from their complex societies.

The first of the three major studies Calhoun proposed for his last cycle of research would take place in a new universe—Universe 33—and focus on what he dubbed "The Ultimate Behavioral Pathology (The Beautiful Ones)." Universe 25 had taught him so many things about behavior and population dynamics, but, of course, he hadn't known what all those things would be beforehand: had he, he would have homed in and gathered more information on certain behaviors. Universe 33 would replicate many aspects of Universe 25, this time in an octagonal configuration which would reduce corner space and provide more vantage points than in the square habitat used in Universe 25. To see if increased habitat complexity might slow population growth and then demise, Universe 33 would be a bit more structurally complex than Universe 25 in that the floor area in each cell was subdivided by a series of partitions that the mice had to climb over to get from the center of the floor to the apartment complexes. Another more significant change was that Calhoun would act as a predator of sorts in Universe 33. He would cull the population periodically to test whether that might slow down the detrimental effects of crowding and overpopulation. Over the course of the first two hundred days, he would allow the population to go from sixteen to two hundred mice, but not more. Then from days 201 to 400, the population could grow up to four hundred individuals, but, again, no more. Finally, from days 401 to 600, Calhoun would allow the population to expand to eight hundred. At that point, he thought virtually all mice would be Beautiful Ones, and the population crash would commence.

Calhoun planned to gather more detailed observations in Universe 33 during "those critical times when inappropriate communication interferes with the maturation of adequate adult behavior." That inappropriate communication included the rejection of young mice by adults, and Calhoun sought to better understand how "contacts with older associates influence the development of mature social behavior," as they set in motion the cycle that ultimately led to the Beautiful Ones becoming so common. He decided that as young mice matured, he would remove a handful of these mice for short periods and put them through a series of tests to measure how they navigate their environment and if they had any "preference for social or complexity stimulation."[4]

The idea was to begin work on Universe 33 in March 1975. Calhoun thought this experiment would take five years to run its course, and his goal was "to understand [how] individual pathology—behavioral and physiological— . . . is influenced by the degree of stability of social organization . . . [and] degree of crowding." As for the experiment's significance for our own species, Calhoun left no doubt: "Overcrowding through a succession of generations is a loss of capacity to execute the more complex behavior for mice," he wrote, particularly behaviors "relating to sex, reproduction, aggression and territoriality." He was becoming convinced the social disconnect that was part and parcel of life as a Beautiful One was the mouse equivalent of autism. Universe 25—where, by the end of the experiment, almost all the mice were Beautiful Ones (Calhoun went so far as to call the prevalence of Beautiful Ones "universal")—was an "animal model of the origin of universal autism," and now Universe 33 would let him dig deeper into that.

Throwing all between-species barriers to the wind, Calhoun warned that what happened in Universe 25, and what he predicted would happen in Universe 33, was a portent of what might come in our own species: "A similar alteration of nervous system function in the human species should make its initial impact in impeding the ability to acquire, create and utilize complex concepts." This "ideationally related pathology"—that is, the potential preponderance of human Beautiful Ones—"could precipitate the extinction of *Homo sapiens*." Understanding how this all unfolded in rodents might, Calhoun argued, help prevent that.[5]

The second prong of Calhoun's "Last Cycle of Research" was a study he labeled "Conceptual Evolution of the Rat." He introduced this study, which would take place in rats living in Universe 34, as the antithesis of what he thought would happen in mouse Universe 33: in Universe 34, he would look not at the unfolding of behavioral pathologies but rather at the possibility of a population "counteract[ing] the deleterious consequences of increased density through conceptual evolution."[6]

Universe 34 would, in fact, be two universes—34A and 34B—and Calhoun estimated the optimal group size in both to be his magic number of twelve. Universe 34A would serve as the control and would begin at that optimum number, with six adult male and six female rats. Each generation, Calhoun would cull the numbers, so that Universe 34A would have no more than twenty-four adults in generation 2, forty-eight in generation 3, ninety-six in generation 4, and then remain at ninety-six for generations 5 and 6. Periodically, he would briefly remove rats from Universe 34A and test them on tasks that would challenge "their capacities to perceive, discriminate [and] associate." When the population reached ninety-six, eight times what Calhoun assumed to be optimal population size for the space in Universe 34A, rats would, he predicted, "exhibit comparable behavioral pathologies" to those of the mice in Universe 25 (and eventually Universe 33), and what's more, they'd do rather poorly on those periodic cognitive tasks.

Universe 34B was where Calhoun intended to foster the conceptual evolution, via cooperation, that would allow rats in that universe to better cope with above-optimal group sizes than their counterparts in Universe 34A. The population in 34B would be allowed to grow just as in Universe 34A, and they, too, would be periodically tested on various cognitive tasks. Each of the rats in Universe 34B would have one of the miniaturized passive resonators implanted under its skin, allowing for rats to be identified as they moved under portals. In generation 2, in order for rats to get water, two would need to press on a lever of a STAW device: but it was more complicated than that, because for water to be dispensed, the portal must register that one rat was male and the other female. In generation 3, not only would there be portals set at "intersex only" at the water dispensers but there would also be portals at the food hoppers, such that, again, two rats needed to press a lever for either rat to get food. The portals at the feeders were set so that rats would only get food if both were "high status" dominant individuals or both were "low status subordinates," so that now complex cooperation was required for both food and water. In generation 4 and all subsequent generations, life was even more complex in Universe 34B. Not only were the portals enforcing different rules of cooperation to get food and water but now portals would be placed at the six areas where nest boxes were in place, and those portals were set so that each of what Calhoun called six "bedroom communities" had to have equal representation of male and female and dominant and subordinate rats.

Rats would be put into Universes 34A and 34B starting in June 1975, and as was the case for the mice in Universe 33, Calhoun anticipated this work would go on for five years (six generations). At the end of those six generations, what he expected to find was that the ninety-six rats in control Universe 34A would be living in a behavioral sink full of rat Beautiful Ones. But the cooperative culture fostered in Universe 34B should, Calhoun predicted, offset the deleterious effects of overpopulation and allow ninety-six rats to live relatively peacefully, meaning they would display the same sorts of behaviors displayed in generation 1, when they were at their optimal group size of twelve. What's more, he predicted that the rats in Universe 34B would also be smarter than their counterparts in Universe 34A, as measured by how well they performed those periodic tasks. "I propose to make the rats in my contrived environment [Universe 34B]," Calhoun wrote, "comparable after five years to apes in their natural environment." With a dramatic flourish, he added, "In essence, I propose to make an ape out of a rat."

One reason that Calhoun thought his metascientific approach was important was that it would generate out-of-the-box, productive new ways of seeing a system. In the case of the rat cooperation/culture experiment, it did just that, as it led him to propose that one of the key components of rat culture was the social networks in which the animals were embedded. Calhoun wrote of "network[s] of overlapping, cooperative subgroups," of "social network[s] [that] may counteract the ill effects of increase in density," and of the way a network "produces greater sensitivity to surrounding conditions and more effective adaptation to them." The study of social networks in nonhumans is one of the most active areas in the field of animal behavior today and is used to study every aspect of animal life, including what they eat, how they protect themselves, who they mate with, the dynamics of parent-offspring relations, power struggles, navigation, play, cooperation, and more. But in 1975, aside from a scant few studies using simple social network thinking in primates, it was barely on the radar.

For Calhoun, this network thinking that he learned in his readings on sociology was critical, for the rat culture experiment was, in large part, an attempt to look at a stripped-down version of the evolution of human culture and the means that we used to avoid sliding into a behavioral sink long ago. His solution was to look at communication networks as a means of avoiding behavioral pathologies. Rats, it struck Calhoun, should be able to form rudimentary networks: they "have much smaller brains than humans," but his "objective [was] simply to encourage them to acquire sufficient culture, compatible with the small brains to offset the pathologies produced by an . . . increase in density above the optimum."[7]

There remained, as ever, the question of the money and labor needed to start these new five-year-long projects. When Calhoun presented his new ideas to NIMH, in an environment where budget cuts for behavioral work were now the norm, to his surprise and delight, "the institutional response to this new direction was amazing." As Calhoun described it, "The NIMH provided a relatively large fund to revamp our research facility and equipment, and enabled me to attract four outstanding young men . . . to work along with me." This new cadre of troops, all postdoctoral associates, required office space of their own. Fortunately, Calhoun's new, unexpected riches allowed for that, so as construction of new universes commenced, contractors also went to work on the remodeled space for humans in Building 112. Looking back today, Norman Slade, one of the new postdoctoral fellows who worked on Calhoun's team, describes that office space as an area that resembled a mouse universe, with water and food placed at a central location and offices, as were the nest boxes in a mouse universe, situated along the sides. "So even if you didn't want to come out and be social for lunch or whatever, everybody had to go to the restroom," Slade says with a smile. "Everybody had to pass through this social space." None of it was happenstance: "[Calhoun] explained the whole dynamics to us."[8]

Sabbatical also provided Calhoun time to spend on his ever-growing database of "population and mental health" that he planned one day to publish as an anthology. By now it had morphed into a more general "environment, adaptation, and population" database akin to an information hub, which Calhoun hoped would be available via the electronic prostheses that one day would make up the human global brain. Calhoun saw this database-one-day-to-be-anthology as serving an important function above and beyond providing raw information that experts from around the world could access. It would, he proposed, provide readers with the opportunity to "encounter different ideas in conjunction before they would find their interfacing through more customary types of publication." Calhoun described a eureka moment in the anthology project: "an amalgam of mind," he called it. "My first image was of several hundred persons stacked on a psychiatrist's couch free associating" on the subject of mental health and population. With all that in mind, Calhoun employed some of his sabbatical energy on "eliciting individuals with demonstrated interdisciplinary competence" to send contributions to the anthology.

Editing an anthology was one thing—writing a book on his decades of experimental work with rats and mice was another matter altogether. Calhoun had turned down numerous solicitations to write such a book, but as he sifted through decades of unpublished data on his rodent work from as far back as his time at JAX, his sabbatical had him thinking that such a book "promise[d] to be a more effective way of treating the past years of effort as a whole, rather than less meaningfully fractionating it into a series of journal articles." Tentative titles he considered at different points in the development of this book included *℞evolution: Prescription for Evolution*, *The Rodent Key to Human Survival*, and *Looking Forward*. During his sabbatical, Calhoun sketched out a very rudimentary outline that noted such a book should have sections on "The Beautiful Ones—Overliving," "Induced Cultural Evolution: Rats to 'Ape,'" and "Prosthetic Brain R & D." That outline also included reminders to cover optimum group size, social velocity, von Foerster's doomsday model, social physics, human evolution, and the compassionate revolution. It would, Calhoun's handwritten notes suggest, cover the work of Darwin, Freud, and a host of others and include a quote from Goethe that captured Calhoun's own long-term perspective: "The fate of the architect is the strangest of all. How often he expends his whole soul, his whole heart and passion, to produce buildings to which he himself may never enter."

While Calhoun was now pondering a book-length treatment of his experimental work in rats and mice, he was still, on occasion, using nontraditional journals to write about his thoughts on both that work as well as on the human population explosion. In the *Psychopharmacology Bulletin*, he published an article—"The Role of Brain Prostheses and Organizational Synergy in Information Metabolism"—that included a brief overview of his ideas on electronic prostheses, appreciative systems, and invisible colleges. In "An Evolutionary Perspective on the Environmental Crisis," a paper he published in the journal *Fields within Fields*, Calhoun preached that the way out of the environmental crisis, within which sat the population crisis, was to understand conceptual spaces, "prosthetic social brains," the compassionate revolution, and how our evolutionary history had shaped us to live in optimal group sizes of a dozen.[9]

In 1975, the same year that Calhoun planned to start the work he described in "Last Cycle of Research," Harvard biologist E. O. Wilson published *Sociobiology: The New Synthesis*. The *New York Times* published four full-length articles on *Sociobiology*, including one months before the book was published and one by Wilson himself. While much of what was written about it praised the book for its integrative approach toward the evolution of social behavior, some social scientists, as well as evolutionarily biologists—including Stephen J. Gould and Richard Lewontin, both colleagues of Wilson's at Harvard—were upset about the book's final chapter on social behavior in humans. In particular, they worried that Wilson's emphasis on how natural selection shapes human behavior suggested that many human behaviors were genetically fixed. "Biological determinist arguments all have a similar form," Lewontin wrote. "A particular model of society is described. It is not surprising that the model of society that turns out to be natural, just and unchangeable bears a remarkable resemblance to the institutions of modern industrial Western society, since the ideologues who produce these models are themselves privileged members of just such societies."

The fact remains, though, that *Sociobiology*, published by Harvard University Press, is the quintessential academic tome. Stuffed with hundreds of figures and an abundance of technical terminology, it runs 697 pages—and oversized pages at that—and, aside from the last chapter, is about the evolution of social behavior in nonhumans, including rats. The mice in Universe 25 had certainly been burning bright in the spotlight in the mid-1970s, but it was Calhoun's Casey Barn rats from the early 1960s that scampered into *Sociobiology*. "John B. Calhoun's famous Norway rat colonies stopped reproducing when the population reached abnormally high densities," Wilson wrote in chapter 4. "Infant mortality reached 80 and 96 percent in two series of experiments. . . . The growth of the young was also retarded in the crowded rat colonies, a phenomenon that is . . . widespread . . . in other kinds of animals." The behavioral sink makes an appearance in that same chapter. Later in *Sociobiology*, Wilson makes note that "bizarre effects were observed" in Calhoun's rats: "some of the rats displayed hyper-sexuality and homosexuality and engaged in cannibalism."

Over the last fifty years, the publication of *Sociobiology* has come to be viewed by many biologists as a watershed event in the study of the evolution of animal behavior: and there, for everyone to read about, are Calhoun's Casey Barn rats sliding into their behavioral sink.[10]

15

MICE TO STAR IN JAPANESE FILMS

Even with the massive media attention it attracted, the reach of E. O. Wilson's *Sociobiology* was primarily biologists, anthropologists, and psychologists. But Calhoun's work was also reaching a completely different audience in architect Barrie Greenbie's book, *Design for Diversity: Planning for Natural Man in the Neo-technic Environment*. Greenbie had spent time visiting Calhoun's lab at the NIH field station and told his readers that Calhoun "devised a number of environments for rats and mice of varying sizes. . . . The mouse-house module might be visualized as a four-story garden apartment . . . fully equipped with automatic vending machines providing all necessities." Knowing his audience was largely architects, Greenbie was quick to note that this "'utopia' [was] not unlike some of the projects that visionary architects of human habitations have dreamed up for our welfare." On pages 36 and 37 of *Design for Diversity* sit striking photos of Calhoun in Universe 25 juxtaposed with a high-rise in the brutalist architectural style of the 1970s. The legend reads: "A Multi-Celled Habitat for Man?"

In *Design for Diversity*, Greenbie wrote of the Beautiful Ones in Universe 25 as well as the rats of the Towson enclosure who had invented the rodent equivalent of the wheel. The comparisons to our species abound: "The fact that innovative behavior in such relatively simple animals appears among the marginal members of society, the 'non-conformists' . . . helps to explain certain well-known characteristics of human innovators," wrote Greenbie. "Artists, writers, musicians, scientists, inventors, explorers, and even certain types of statesmen, tend to be 'loners.'"[1]

About the time that *Sociobiology* and *Design for Diversity* taught readers about population growth and demise in mouse and rat universes, Calhoun, who remained an attraction on the academic lecture circuit, presented a lecture on "What Population Studies of Mice Tell Us about the Future" to a joint meeting of the Smithsonian Associates and the National Audubon Society. Not long after that Calhoun was asked to talk to an undergraduate class on the Natural History of Man at Johns Hopkins University, where he spoke on "Essences of Existence and Evolution: Studies on the Lives of Rodents and Humans."

The media coverage of Calhoun's work was no longer what it had been through 1973, when "Death Squared" was published, but there were still pockets of interest. In 1976, the *Sun Magazine* ran a huge six-page spread, "Of Mice and Men and Escaping the Ultimate Pathology," with photos that spanned from the days of the Towson enclosure work to the earliest stages of Universe 33. The following year, in a story about rats that had survived early thermonuclear tests on Enewetak Atoll, the *New York Times* mentioned Calhoun's work from way back in the days of the Rodent Ecology Project in Baltimore.

As ever, Calhoun was busy writing, although mostly prospectuses, reviews, and book chapters rather than papers packed with *new* data. In "A Scientific Quest for a Path to the Future," an article he published for *Populi: Journal of the United Nations Fund for Population Activities*, Calhoun gave an overview of his work on mouse and rat population growth and demise, and he discussed his ideas on electronic prostheses and a global brain. In an autobiographical book chapter titled "Looking Backward from 'The Beautiful Ones,'" Calhoun walked readers through his early days in Tennessee to the commencement of mouse population dynamics in Universe 33 and rat cooperation in Universe 34. The editor's introduction to that chapter prepares the reader for what is to follow. "Early in his career [Calhoun] skipped around from place to place, mostly in untenured, low-paying positions," William Klemm wrote. "In almost scatter-brained fashion, he seems to have flitted from seemingly unrelated idea to idea, experience to experience. Yet in the process, maybe because of the process, he created many new insights about the behavior of animals and humans."

More than anything else, though, for John Calhoun, the second half of the 1970s was all about collecting data on both the mice in Universe 33 and the rats in Universe 34.[2]

Kathy Kerr of the Bowen Family Center near Georgetown University in Washington, DC, was one of the troops in Calhoun's army of Universe 33 data-takers. Kerr had heard Calhoun lecture at the Bowen Center many times, and when Calhoun circulated an advertisement looking for volunteers in 1978, Kerr jumped at the chance. That summer, one day a week, Kerr and a friend drove out to Building 112 at the NIH field station. The first step was to be trained, which involved learning to read the numbers that were painted on the backs of the mice. Next, Kerr familiarized herself with a mouse ethogram—a list of the key behaviors—so she could write down what a mouse was doing as she watched. Once that was done, she started gathering data. "It's cold [and] you walk into this big, circular room, and in the middle of it is the universe and around [it] is a platform," says Kerr, describing Universe 33. "You can stand on the platform and look down into the universe." She watched the mice from a fixed point for a few minutes then moved to a new point, circulating around the room and collecting data on feeding, drinking, and fighting, who was giving birth in nests, who was giving birth on the floor, who was giving birth out in the open, how good of a mother a female was, and more. In time she got to know the mice personally: "To this day, I remember this male mouse . . . number 103. The population density was [increasing] . . . and yet he maintained competent social behavior. And I was like, 'Oh my God, I'm impressed with you, little guy. I want to be you in my lifetime.'" And of course, Kerr got to know the Beautiful Ones, who, she says, "did nothing and who had no effective social behavior skills. But they were pretty." Once she was finished doing observations in one cell, she'd then move on to the next cell and do the same thing.

The only break all day came at lunch when Kerr would join the whole Universe 33 army, often including Calhoun, for a quick meal. Some days Calhoun talked about his ideas on a human global brain and what the mice might teach us about that. Kerr recalls Calhoun eating a sandwich while discussing "the idea of the colony as a brain. . . . The [mouse] colony as a wasp colony or a superorganism. . . . The sum of the universe was more than the individuals in the universe." Looking back today, she now thinks he had "the world wide web inside his head."[3]

Calhoun's NIMH progress reports from the second half of the 1970s, as well as periodic updates written in NIH newsletters, paint a picture of how the experiments in Universes 33 and 34 were coming along. And things were not going well, at least as far as life for the mice in Universe 33. Despite Calhoun's periodic predatory-like culling to try to slow population growth, by 1978 the mice in Universe 33 appeared to be on the same path to extinction as their predecessors in Universe 25. By day 200, there were already hints of a behavioral sink starting to form. When Calhoun looked at where mice were during a twenty-four-hour cycle, they showed a very strong preference to spend their time in one of the sixteen cells. They also took up residence in half of the apartment blocks in their octagonal universe, leaving the other apartment blocks relatively uninhabited. By day 400, more and more evidence of a behavioral sink had emerged: the Beautiful Ones were becoming more common. Still, Calhoun had underestimated how quickly the population would be made up largely of Beautiful Ones, on an unavoidable path to extinction. He had hypothesized that would happen in generation 5 at a population of eight hundred mice, but in his progress report for 1978 he revised that estimate: "It now looks as though we will have to follow the history of this population for another 400 to 600 days until the 6th generation is in adulthood and the total population stands at near 1600."

By 1981, Calhoun had, in fact, continued the experiment not just to generation 6 but to generation 8. And between generations 5 and 8, things for the inhabitants of Universe 33 took a very sharp turn for the worse on many fronts. While "signs of behavioral disintegration and social instability" had begun in generation 4, they were accelerating rapidly. Dominant males, with high social velocity, were always relatively aggressive, but now females and subordinate males were ramping up their aggression as well. Females became progressively worse at building nests that could sustain young, and more and more were giving birth in public spaces—the floor of the octagon—rather than in their apartments. Mortality was very high for pups born on the floor, but it was also getting high in nesting boxes. Across all sixteen cells, nesting boxes, and floor space, half the pups lived no more than a few days. By generation 7 the population in Universe 33 had reached sixteen hundred, at which point it "entered a phase of temporary numerical stability where the few surviving new-born members just replace those from earlier generations that die." Calhoun still assumed twelve was the optimal number of mice (and just about everything else) in a group, and by generation 7 only 2 percent of the mice—almost all of whom lived in apartments—were found in groups around that size, with the vast majority of the others spending virtually all their time on the public floor space: these were the Beautiful Ones, who Calhoun described as suffering from "extreme [social] withdrawal."

By generation 8, the behavioral sink and the overabundance of Beautiful Ones seemed to have sealed the fate of the mice in Universe 33. The population crashed from sixteen hundred to eight hundred, and, worse yet, no females from generation 7—indeed no remaining females from generations 5 or 6 either—produced any progeny that survived. As far as what Calhoun thought all that meant for our own species, he noted that "evidence from inquiries on humans indicates that humans, like mice, exhibit behavioral and social pathology when the number and rate of contacts increase beyond a fixed level," and that meant that "the evolutionary transition period of 1975 to 2175, with its doubling of human population and the drastic change in character of roles demanded, presents a megacrisis period very comparable to our mouse population."[4]

As mouse society was unraveling in Universe 33, the rats next door in Universe 34 were shedding new light on culture, cooperation, and population growth. *ADAMHA News* (from the Alcohol, Drug Abuse, and Mental Health Administration), one of NIH's newsletters, published a story touting "an unusual 'brainstorming' session" held at the NIH field station. Calhoun, the newsletter told its readers, had convened a two-day symposium there that brought together sixty experts from evolutionary biology, ecology, architecture, public health, psychology, and psychiatry "to discuss a new project which will test the hypothesis that the effects of overcrowding can be overcome." Symposium members paid close attention as Calhoun explained the setups in Universe 34A (the control universe) and 34B and argued that the need to cooperate in Universe 34B "will give the rats further social organization beyond what nature would allow," which would counter the effects of rampant population growth. Attendees were excited by what they heard and "expressed the view that studies of this type are important now because the world population is nearing a critical period during which crucial decisions affecting the future of man must be made."

In subsequent NIH newsletters, the denizens of one of the largest aggregations of scientists in the United States learned more details: rats in Universe 34B were "given the advantage of becoming members of small subgroups and developing cooperative social roles." Calhoun hoped "these arrangements [would] maintain the animals emotionally unscathed despite [an] increase in density." But things would be different in universe 34A: Calhoun predicted that the fate of those rats would be the same as the Beautiful Ones. "We're reducing the whole evolution of culture to a simple paradigm," Calhoun told the NIH reporter. "If you can design environments to guide social relationships, you can give each individual an optimal number of contacts."

Calhoun's environmental designing seemed to be working. Rats in Universe 34B, where they needed to cooperate to do just about everything, were up to the task(s). They not only learned to cooperate but, over the course of generations, they showed a marked reduction of aggression, compared to what was taking place in Universe 34A. Rats in Universe 34B were also able to maintain their normal body weights far better than the control rats in Universe 34A. Most importantly, with respect to population growth, in Universe 34B the rats developed "[new] modes of reproductive control," and so Calhoun concluded that the "acquisition of cooperative behavior reduces crowding related stress." Reproductive control occurred not by differences in the number of pups females gave birth to in Universe 34A versus 34B but rather by a marked decrease in the rate of conception in Universe 34B. Calhoun interpreted all this as evidence that enforced cooperation in Universe 34B slowed population growth.

An interesting twist that Calhoun did not expect was that every time he added a new cooperative task (water, food, etc.), there was a period of social instability, during which the rats in Universe 34B seemed to be figuring out the new cooperative challenges they now faced. "This leads to a generalization," Calhoun wrote in one of his many NIH progress reports. "Maximal profiting from introduction of new social roles must allow for a relaxation interval of time, prior to introduction of another new cooperative social role."

Every NIH progress report Calhoun submitted required the lead researcher to include a paragraph on "Significance to Biomedical Research and the Program of the Institute." "Humans are exposed to environments which are increasingly becoming more variable, complex, and changing," Calhoun wrote there. "This historical trend has increased our capacity to create and utilize ideas as well as to cope with change." That was all fine and good, but Calhoun noted, "The current rate of change introduces undesirable stress on individuals and produces broader instabilities in society. Our research with rats suggests that promotion of adoption of more types of cooperative social roles can ameliorate stressful consequences of physical and social environmental change."[5]

Initially, Calhoun had planned to let Universe 33 and Universe 34 run for five years, but as time went on, he decided to allow work in both to go on for an extra year or two. And because Calhoun felt like he needed to have those experiments run their full course before publishing results, he rarely submitted any new rodent population dynamic papers for publication. But his work on the Casey Barn rats and the mice of Universe 25 was being cited often not only in journals from a wide array of disciplines and covered in newspapers (though not as often as in the early 1970s) but also in dissertations and major textbooks.

Dozens of dissertations in the mid to late 1970s wrote of Calhoun's work on population dynamics in mice and rats. These dissertations covered a dizzying array of topics ranging from "The Social Behaviour of Adult Rats Undernourished in Early Life" and "Transport of Young in the Norway Rat" to "Some Effects of School Building Renovation on Pupils' Attitudes and Behaviors in Selected Junior High Schools" and "The Effects of Population Density, Room Size and Group Size on Human Intellectual Task Performance and Emotional Reactivity," which emphasized that "Calhoun's experiments are summarized at length since they provide the material from which a vast number of conceptions about the effects of 'crowding' have originated."

In their textbook, *Animal Behavior: Concepts, Processes, and Methods*, Lee Drickamer and Stephen Vessey devoted two pages to Calhoun's work alongside a large photo of Calhoun standing in the middle of the Universe 25. "Calhoun observed several aberrant kinds of behavior in male rats and mice in his 'universes,'" undergraduate and graduate students interested in animal behavior learned. The text included a discussion of "pansexual males," "probers," and Beautiful Ones, the last of which "have flawless coats with no scars and walk about the pen with impunity, being ignored by territorial males and even nesting females." The Beautiful Ones, be they rats or mice, Drickamer and Vessey wrote, were "in a state of physical and mental immaturity."

In the second edition of her textbook, *Introduction to Psychology*, Linda Davidoff discussed Calhoun's work. Davidoff's publisher contacted Calhoun for a photo, and this time he sent along a picture of octagonal Universe 33, where he and his colleagues were looking down over the mice therein: in the legend to that figure, Davidoff labeled Calhoun a "medical researcher." In the text Davidoff noted that "mice in an earlier study [Universe 25] multiplied naturally over the course of about 1,600 days." Yet despite abundant food and water, no predators and no "inclement weather," she tells young psychology students, "competition became intense after a few generations. . . . [There were] breakdowns in parental care, territorial defense, and normal social maturation." The Beautiful Ones also make an appearance in Davidoff's textbook in the form of mice "who learned little about social conduct: courtship, parenting, and aggression."

Television also played a role in Calhoun's work gaining international attention. Jun-Ichiro Takeda, president of the Cine-Science Company in Japan, wrote Calhoun to tell him of a series of shows on human psychology that Takeda's company was producing in both Japanese and English. Takeda was "deeply impressed" after reading some of Calhoun's work and wrote that "we should like to emphasize that it is very important to film [these] studies to introduce a model of life on earth in limited space and circumstances." Takeda requested that one of his camera crews be granted access to Calhoun's lab to film the rodent universes so Takeda could include Calhoun's studies on crowding, overpopulation, and population crashes in his series. Calhoun was more than happy to oblige but told Takeda that his crew would need to be patient once there: "I have worked with camera crews from all over the world and I am sure that you could obtain sequences here that will fit well into the theme you will attempt to integrate; however there are conditions," he wrote Takeda. He next explained that it often takes a film crew hours to get the lighting just right for filming, and that "animals do not follow script, you just have to wait until [the appropriate] social behavior . . . actually occurs . . . even 25 to 75 seconds of the most pertinent shots may take one to two hours." All this was fine from Takeda's end, and soon a crew was in the Calhoun lab filming. Not long after that, an NIH newsletter had an article touting not Calhoun but his subjects: "Calhoun Mice to 'Star' in Japanese Films."

It wasn't only Japanese film crews who were interested in visiting Calhoun's universes. America's newsman, Walter Cronkite, was developing a documentary series called *Universe* for CBS News, and the producers contacted Calhoun to learn more about his work for a possible visit to the NIH field station. Calhoun immediately sent copies of thirteen of his publications, including the *Scientific American* Casey Barn rats paper and "Death Squared," the *Proceedings of the Royal Society of Medicine* article on the mice of Universe 25. In the cover letter that went along with that package, Calhoun enclosed an overview of his current projects, including the brain prosthesis work, and he couldn't resist sharing a story about something that happened after CBS had initially contacted him. He told one of the producers of *Universe*, "One of my colleagues remarked, 'well, isn't that appropriate, universe—that's what we work with all the time.' [My colleague] was referring to the fact that we call our large experimental habitats universes, e.g., Universe 33 for studying social impact of overcrowding in mice and Universe 34 for simulating the origins of human cultural evolution—this latter with rats." Calhoun and the producers of *Universe* exchanged a few more letters, but before anything came of a possible visit to Building 112, though *Universe* was widely acclaimed for content, the show was canceled for poor ratings, and so Walter Cronkite never did get to meet the rats and mice in Building 112.

As always, Calhoun was giving lectures. In 1980, he was back at the Smithsonian to speak in a new series they were organizing called "Creating the City: Design Dialogues." In the advertisement for his talk, "How Do We See Cities," Calhoun was listed as an "ecopsychologist . . . [and] leading theorist on crowding." "How Do We See Cities" was a sprawling talk that opened with Calhoun telling his audience that "the study of the consequences of various choices open to human societies and individuals should have the explicit aim of contributing to a spectrum of values that is both scientifically informed and humanly applied." A moment later, he added that "certain features of animal behavior have suggested promising hypotheses about man." Harkening back to social physics, Calhoun noted that individuals and groups in social systems, human or otherwise, are akin to molecules moving about in the universe, strongly influenced by information and communication, which led to a discussion of the electronic prostheses he envisioned coming to pass. From there, he outlined his idea on human cultural evolution through revolutions, the next of which would, hopefully, focus on appreciative systems and zero population growth.

In his attempts to tie together human and nonhuman behaviors to understand group living, Calhoun never explicitly discussed his work on population dynamics with rats during his Smithsonian lecture, and he only mentioned his mouse work in passing, referring to "utopia-turned-hell [mouse] universe(s)"—and for a lecture on a series on the city, there was remarkably little discussion at all on cities. Calhoun did argue that his work on the stages of population growth and demise suggested we need to realize that "environments which enhance human adaptation [for] cities too have developmental stages." How to do that was the problem. Calhoun asked his audience, "Should we look at the components of our cities as experiments of Man and think in terms of survival of the fittest?" If so, fitness needed to be measured in terms of enhancing creativity. This would take time and lots of work, but Calhoun was optimistic: "We are on the leading edge here of an exciting and meaningful dimension and mission for our cities. Let us widen that edge so that it dominates the future."[6]

In addition to work in Universe 33 and Universe 34, Calhoun began a spin-off experiment on rat cooperation, culture, and population growth. He was certain that what was happening in Universes 34A and 34B supported his hypothesis that cooperation tempered population growth. Now he wanted even more details on how, so he constructed yet another universe. Again, rats would have to cooperate to gain access to food and water, but this new universe—Universe 35—differed from Universe 34 because one-half of it was painted light and the other half dark. Rats are nocturnal, and Calhoun assumed dominant males would fight for, and eventually set up residence in, the dark side, which is just what they did. Many females on both sides got pregnant and reared offspring, but, not surprisingly, females who nested with dominant males on the dark side of Universe 35 had better overall reproductive success than females on the light side. At the level of the entire universe, by creating their own light and dark subdivisions, rats, Calhoun believed, had partitioned Universe 35 in such a way as to allow the total population to increase more than it would have otherwise. What Calhoun didn't expect was that some rats had their own means for enforcing cooperation: a rat slavery system of sorts emerged. But not everywhere.

Calhoun assumed that the subordinate rats, who would have to settle for life on the lighter side of Universe 35, would develop "a safety factor for adapting to changed circumstances." This prediction was based in part on "past studies show[ing] that members of such subgroups have a higher probability of developing both deviant and novel behaviors, some of which provide them with a readiness to accommodate to changed circumstances, and thus enhance species survival." The rats on the lighter side did not disappoint, as they settled into a routine of cooperating for both food and water. Things were different on the dark side of the Universe. Some dominant rats there cooperated with one another but others "developed a slave producing behavior," wrote Calhoun. "A dominant rat would chase a subordinate into one tunnel of an apparatus [to get water or food] and force it to stay there." On the light side of Universe 35, rats had learned to cooperate and kept their population under control. On the dark side, though more babies were being born, some rats had, Calhoun argued, sunk to slavery.[7]

As life, death, cooperation, behavioral sinks, and slavery were unfolding in his mouse and rat universes, Calhoun continued to think about his electronic prosthesis as a means for saving us from ourselves. Ultimately, he felt the anthology he was working on about environment, population, and mental health would serve as a starting point, a small part of what eventually would be the giant worldwide brain linked by his electronic prosthesis. In his NIH progress report for his "Conceptual Prosthesis of the Brain" grant, Calhoun told his higher-ups that "to the extent that this expectation is realized it will contribute to the continuation of the evolutionary process by enabling more rapid and more complex interrelationships among ideas than is possible alone from biological brain processes and the process of interpersonal communication developed through culture."

To better explain how this would all happen, Calhoun borrowed the concept of negentropy from the field of information theory. As opposed to entropy, a measure of the disorder and randomness in a system, negentropy is a measure of order and complexity. Calhoun hypothesized one outcome of his electronic prosthesis would be that "the negentropy level of this part of the universe will be increased as a result of greater amount, diversity, and complexity of information having survival utility." Negentropy was even more fundamental than that: "'Tradition,' in the biological brain," Calhoun wrote, will be replaced by "'negentropy enhancement' . . . in the multi-person prosthetic brain." When asked by the Northern California chapter of the American Academy of Pediatrics to present a series of six lectures on a fall 1981 Mississippi River Cruise the chapter had organized for its members, Calhoun titled his lectures "Toward a Negentropic Model of Being and Becoming." As Northern California pediatricians and their families dined on food from the bayou, Calhoun enlightened them about not only negentropy and prosthetic brains but also behavioral sinks and optimal group size in mice and rats.[8]

NIH was paying close attention to what was going on in Calhoun's universes, and it was doing more than just funding his studies and publishing newsletter stories about his work. In 1981, the Alcohol, Drug Abuse, and Mental Health Administration (ADAMHA)—an umbrella group that included NIMH—honored Calhoun with their Award for

Meritorious Achievement for his "innovative research on environment on behavior, and the impacts of crowding on animal and human populations."

It should have been a day of celebration for Calhoun: it turned out to be anything but. The awards were presented by William Mayer, who, just a month earlier, had been appointed ADAMHA's new director. During informal remarks before the presentation, Calhoun recalled Mayer telling those gathered that under his directorship, "The NIMH is drugs. Period." For someone who had spent a career studying behavior in nonhumans, and virtually never incorporated drugs into that work, it was an omen of ill winds coming.[9]

16

THE RODENT KEY TO HUMAN SURVIVAL

Despite what Calhoun viewed as the perilous new path NIMH was heading down in 1981, for the next two years he had the funds he needed to complete the mouse work in Universe 33 and the rat work on cooperation and population control in Universes 34A and 34B. As the experiments in those universes were entering their final stages, newspaper and magazine articles about his work had essentially come to a halt—largely because Calhoun simply was not publishing any new results from Universes 33 and 34. But, *Encyclopaedia Britannica* still saw fit to send a film crew to Calhoun's lab to get some footage for an educational documentary called *Animal Populations: Nature's Checks and Balances.*

The entire documentary was twenty-two minutes long, and Calhoun's work made up more than a third of that. The narrator opened by noting that "in addition to his renowned research papers, [Calhoun] has recorded some of his observations on film," at which point a video clip from Universe 25 is spliced in, and the mice there are described as living in "a utopian paradise." With haunting music playing gently in the background, viewers learned of the Beautiful Ones, "who never involved themselves with others, engaged in sex, nor would they fight." The narrator continued, "All appeared as beautiful exhibits of the species with keen eyes and healthy well-kept bodies." Viewers learned that besides their sex lives, the Beautiful Ones had another problem: "Though they looked intelligent, they were, in fact, very stupid. . . . Eventually the entire mouse population perished." Following the tale of Universe 25, a very brief description was given of rat work led by James Hill in the Calhoun lab, including an up-close-and-personal shot of the surgical procedure used to implant the glass-encapsulated coils for tracking rats when they passed under portals.

As the Calhoun component of *Animal Populations* draws to a close, the backdrop shifts to the congested streets in New York City's Times Square, and the narrator says, in a rather ominous tone, "Though these studies used animals . . . the findings are being closely compared to our own human population."

When *Encyclopaedia Britannica* sent out advertisement packets for *Animal Populations*, no part of the film garnered more space than that dealing with Calhoun's universes. A photo of Calhoun holding a rat by its tail was added for good measure, and among discussion prompts provided for educators to use with students after they had watched the film was this: "Discuss the stages of social evolution demonstrated in Dr. Calhoun's experiments with the mouse utopia."[1]

Shortly before *Encyclopaedia Britannica* visited, Calhoun was invited to be part of a symposium on "The Human Knowledge Process: An Evolutionary Dilemma" at the 1982 American Association for the Advancement of Science meeting in Washington, DC. It had been nearly a decade and a half since he had first spoken of his ideas about electronic prostheses at the 1968 AAAS meeting at Berkeley, and it was time for an update. Calhoun decided his lecture—which he titled "The Transitional Phase in Knowledge"—would provide that update.

Solving the population problem, Calhoun told an audience of the best scientists in the United States, required revolutionary new ways of thinking. The problem was that we don't have powerful enough brains to solve it. Regardless of how much smarter we are than other species, our "biological brain reached maximum functional potential 43,000 years ago," Calhoun told his audience. We came up with ways to handle that, Calhoun went on, but they had run their course as "the network of interpersonal communications enhanced by mostly non-electronic means (prostheses) of preserving and transmitting information [had] essentially reached maximum functional capacity." What to do about that, Calhoun argued, was *the* pressing question of the time. Biological brains and nonelectronic prostheses, such as (nonelectronic) social networks, "represented the last two leading edges of evolution that have promoted negentropy increase." Each was "concerned with thought and entropy," but each had run its course.

As a surprised audience of scientists, not used to what many would think of as wild philosophizing, listened, Calhoun told them what needed to happen to save us from destroying ourselves: "[The] next phase shift in evolution should utilize computer-like devices and interlink systems of them to simulate biological brain function to manipulate more effectively thought products of the human brain and process them for return use in human reflection." If that wasn't radical enough, *how* it would happen remedied that: "During the coming era the already apparent mutualism between humans and information metabolizing machines will become more pronounced," Calhoun declared. "Machines will more and more represent enlargements of human being, identity, and capacity rather than independent forms of life. Together they will form a higher order of life, increasing negentropy on earth through a natural process of affective and intellectual reactivity."

As for Calhoun's own contribution to how to conceptualize a global prosthesis before the technology existed to create one, in 1983—nearly twenty years after he started gathering and analyzing data for an anthology that began with a focus on populations and mental health—Calhoun published an edited anthology with a much broader scope, titled *Environment and Populations: Problems of Adaptation*, and sub-subtitled *An Experimental Book Integrating Statements by 162 Authors*. He saw this anthology as a paper version of what would one day be a tiny part of a giant global brain linking people and ideas via electronic prostheses.

Environment and Populations was a book, Calhoun told his readers, "In the customary sense of focusing on some circumscribed set of concerns." But, more than that, it was also an "experiment, based on ideas about brain functioning and information processing." Calhoun had asked each of the 162 authors to send him a short entry "about an important researchable problem within a broad domain suggested by the words adaptation, population and environment." He selected those three words, he explained, because of "a belief of mine that history and evolution have entered a period of drastic change, megacrisis, and opportunity—the likes of which this earth has never experienced before and will not for a very long experience again."

Over the decades in which he had been working on this project, Calhoun had tapped into a huge National Library of Medicine database of more than a quarter of a million published articles. He selected more than five thousand of these papers to focus on and then culled the list down to two thousand. Combined with other work he had done, Calhoun said that the information he gathered was "so rich that my mind was incapable of retaining elements that might be interfaced or integrated for consensus or creative purpose." That required some sort of innovative solution, and the one he came up with was "to identify generically 'equivalent' words as suggested by in-context usage." He had come up with "several thousand such pairs."

Using those generically equivalent words, Calhoun then analyzed the short entries (typically two to three pages) of each of the 162 contributors. He was looking for ways to connect what, on the surface, appeared to be disparate trains of thought from medical doctors, psychologists, ecologists, chemists, sociologists, mathematicians, demographers, city planners, computer scientists, architects, economists, anthropologists, and more, which included Edward Hall (proxemics), fellow Space Cadet Leonard Duhl, Margaret Mead, and Stanley Milgram, whose "small world" research from the late 1960s suggested any two humans can be linked using five intermediate acquaintances.

To do this analysis, Calhoun turned to the social network approach he had used in his work on rat cooperation and population growth. He clustered word equivalents until he had a social network with thirteen hubs—hotspots through which much of the information in 162 contributions could be mapped. At the center of that network sat "negentropy enhancement," which Calhoun annotated on the network diagram as "the core of evolution."

In a social network, information flows both directly and indirectly. In Calhoun's network, "acquired characteristics" directly enhance negentropy. But acquired characteristics also indirectly affect negentropy, as they enhance social system evolution, which in turn increases negentropy: the caveat here is that increased social system evolution can create stress, which can negatively impact negentropy. Networks, including this one, can also contain feedback loops: in Calhoun's network, negentropy positively affects ecosystem complexity, which then cycles through social system evolution to increase negentropy.

Putting aside all the network complexity, two things about this analysis were particularly important to Calhoun. For one, his idea

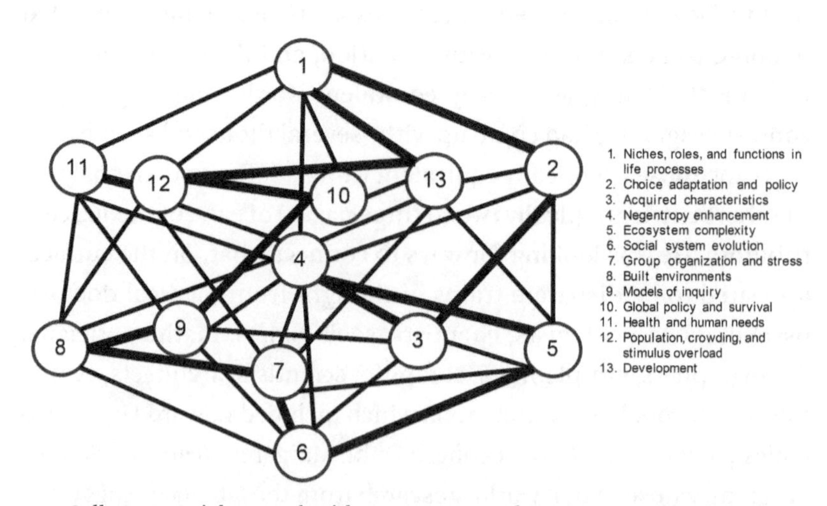

Calhoun's social network with "negentropy enhancement" at the center. Redrawn and adapted by author from J. B. Calhoun, *Environment and Populations: Problems of Adaptation: An Experimental Book Integrating Statements by 162 Authors* (New York: Praeger Publishing Co., 1983).

that increased negentropy was critical for the survival of our species was confirmed when he made the move from intuition to social network analysis. Equally, if not more, important was how these findings would shape his future work. He intended to continue to do the sort of research that had led to *Environment and Populations*, and he began referring to that new work as his gedankenexperiment (German for "thought experiment"), a term Albert Einstein used to describe conceptual experiments when physical experiments are not yet possible. Calhoun was convinced that when computer power was much greater, more sophisticated analysis would, one day, make global prostheses a reality. At that point, "one could then write a computer program . . . to assist a person, [to move] from personal idiosyncratic bias, to produc[ing] a synthesis in an orderly manner." Calhoun thought that "such data bases could serve as an 'early warning' Alerting System for recognizing the existence of emerging problems and their general character." The potential was great. But Calhoun knew that one always needed to be on the lookout for Big Brother: "Extreme care needs to be taken," he wrote, "to assure wide access to such data bases in order to prevent their misuse by persons and institutions with narrow selfish interests."[2]

By the time *Environment and Populations: Problems of Adaptation* was published, the experiments in Universe 33 and Universe 34 were over. Recall that by generation 8, Universe 33 was full of Beautiful Ones, the population had plummeted from sixteen hundred to eight hundred, and no females produced any progeny that survived past a few days. It wasn't long before the population went extinct. But before it did, as he was wont to do, Calhoun removed a few mice and placed them into much smaller groups to see how they would fare. When he had done that with rats in the Casey Barn, they were never able to recover any semblance of rat normality. When the same procedure was used on mice from Universe 25, they fared a bit better, but only a bit. When placed in uncrowded environments, they suffered the same fate as the Casey Barn rats—but when Calhoun placed them into what he called "solitary confinement," Universe 25 mice recovered normal behavior. The mice in Universe 33 were different: when generation 8 mice were placed in small groups, they returned to their normal mouse selves, to the extent that females successfully reared healthy offspring. Of course, Calhoun was curious as to why mice in Universe 33, but not Universe 25, were able to turn things around when they were removed from overcrowded, doomed populations, but with a sample size of just two universes, his explanatory power was limited. His hunch was that the culling he had done in the population in Universe 33 might have played a role.

Years after the completion of the work in Universe 33, Calhoun was moved to write a short ode to the Beautiful Ones, who had taught him so much about behavior, population growth, and population demise. "Such rephrasing of findings or conclusions," Calhoun said, "often helps me understand better where I have been or might go." One day while waiting in the doctor's office, "in between snatches of reading Tom Robbins' *Jitterbug Perfume*," he waxed poetic about the Beautiful Ones:

> . . . all roles, all tradition, just forgot
> No visions new to weigh us down, just floating
> Thus I today will pass unseen

The experiments in Universes 34A and 34B, designed to examine the effect of enforced cooperation on population growth, came to an end about the same time as those in Universe 33. Calhoun knew females in 34B had lower conception rates. The final analysis suggested why: mothers were spending so much of their time learning to cooperate, and then cooperating with others, that they just were not getting pregnant as often as those in 34A, which kept population growth in 34B in check. Indeed, it did more than that, as offspring that did survive in 34B were healthier than those that survived in 34A, the control universe with no cooperation. "The words relaxed, tolerant, empathetic, altruistic and compassionate," Calhoun noted, "best conveyed the quality of overall behavior of the experimental rats derived from learning . . . cooperative tasks."

In his final NIH progress report on Universe 34, Calhoun left no doubt as to what he took the human implications to be: "Engaging in cooperative behavior is the hallmark of culture and civilized life," he wrote. "Studies on rats should help us understand the origin of the greater than exponential rate of increase in cultural evolution among humans which began some 40 millennia ago." Looking to the future, rather than the past, Calhoun was absolutely convinced that understanding this process was the key to saving ourselves from death by overpopulation: "The process of acquiring and maintaining cooperative behavior," Calhoun proclaimed, "should become particularly relevant during the 1975–2175 evolutionary transition when cooperating in enhancing information metabolism is likely to become much more important."[3]

For all intents and purposes, the completion of the work in Universes 33 and 34 marked the end of Calhoun's experimental work on rodent population dynamics. Next would come data analysis and synthesis. The higher-ups at NIH seemed pleased with Calhoun's final round of experiments and displayed a patience that is not all that common in the world of science administrators. "From inception to conclusion . . . the stages of Calhoun's work on animal populations might be compared to those of the reproductive cycle of invertebrate groups involving an egg stage, larval stage, pupa stage and adult stage," said one NIH evaluator, who focused strictly on Calhoun's studies of the population biology of rats and mice. In a rare pat on their own back by administrators, another evaluator, who, again, focused solely on the work in rodents with no mention of Calhoun's ideas on human population dynamics, wrote that the "Administration of the Department of Intramural Research draws great credit to itself for the courage and wisdom it has expressed in providing adequate support for Calhoun's research, despite the long delay in presenting results." A third evaluator remarked that "the nature of [Calhoun's] investigation exemplifies the importance in the present times of the availability of government resources to support long term research that goes beyond the 'quick fix' and gives heed to the long cyclic changes affecting future generations." In this evaluation not only is the rodent work reviewed and praised but the evaluator also noted that the "relevance of such studies to problems facing humanity today needs no emphasis" (though, without elaborating on why). Unlike the other evaluators, this one mentioned Calhoun's edited volume, *Environment and Population*, writing, "[It] provides remedial recommendations in regard to public health. . . . Calhoun foresees one means by which a society can look down the road and take action to steer around potential catastrophes."[4]

During his sabbatical in 1974, Calhoun had rekindled the idea of writing a book on his experimental work on population growth and demise, but what with all the studies he began shortly thereafter, when the sabbatical ended, the idea of a book returned to the back burner for nearly a decade, until he was approached by his old colleague, editor Anders Richter. It had been more than two decades since Richter, then an editor at the University of Chicago Press, had read of the Casey Barn rats in *Scientific American* and contacted Calhoun to inquire whether he might be interested in writing a book. Richter and Calhoun had kept in touch over the years, and in early January of 1983, Richter, who had moved to Johns Hopkins University Press, once again asked about a book that would provide readers with an overview of Calhoun's nearly four decades of work on population growth and demise, with a particular emphasis on the work he had done over the last ten years. Richter soon set up a visit to Calhoun's lab, where he told Calhoun that "the Hopkins Press would like to help you get a Nobel Prize."

This time, the timing was right. Calhoun had been holding off publishing his work on Universes 33 and 34 until that work was complete—which it now was—and Johns Hopkins University Press was a well-respected publisher in the world of academia. Calhoun saw this as an opportunity to write his magnum opus and sent Richter a package containing "background material for a proposed book, 'The Rodent Key to Human Survival.'" Soon Richter and Calhoun were hashing out the details of a contract. But, as Calhoun was an employee of the federal government, writing a book was not as straightforward as it would have been otherwise, because Calhoun needed the approval of NIH to proceed. For that approval, NIH required a detailed external peer review process that involved written assessments, both as Calhoun completed different sections of the book and once a draft of the book was complete. NIH agreed, in principle, to budget of $10,000 to pay seven external readers to review the book from start to finish.

Before Calhoun could take pen to paper to begin writing *The Rodent Key to Human Survival*, two hurdles were thrown in his path. He had been suffering from angina pains, and in December 1982, he suffered a mild heart attack; in April of 1983, he underwent heart bypass surgery. Two months later, as he was recovering, Calhoun received news from the powers-that-be at NIH that he would have to vacate his nearly nine-thousand-square-foot lab in Building 112. Steve Suomi, a primatologist studying the effect of genetic and social factors on rhesus macaque development at the University of Wisconsin, had been hired by the National Institute of Child Health and Development, and the decision was made that he'd take over Calhoun's lab. Decades earlier, it had been Suomi who worked as an undergraduate assistant for Paul Ehrlich, gathering data on nonhumans, and who even earlier than that, as a high school student, had been inspired by Calhoun's 1962 *Scientific American* article—now his macaques would take over the space that had long been home to Calhoun's rats, mice, and human research team. Suomi wasn't pleased that moving into his new lab would displace Calhoun, he says today, but he had little choice in the matter.

With his experimental work on rats and mice ended, Calhoun knew that *eventually* his lab would be downsized, but NIH's decision came out of the blue and seemed rather cold given his recent bypass surgery. He was told he would need to move to Building 110 at the NIH field station and that he and the team working with him on analyzing results would be housed in nine hundred square feet: a 90 percent reduction from what they had in Building 112. Ever since William Mayer had said "The NIMH is drugs. Period," Calhoun had sensed that the days in which NIH would support the sort of work he was doing were numbered. Now he was convinced: "An apparent fixed decision made that Calhoun would be terminated by Oct. 1, 1986," Calhoun wrote, as he kept track of unfolding events. How he came up with that date, he didn't say.

As with all things bureaucratic at NIH, the process of moving the lab was agonizingly slow: it took eight months to complete the move from Building 112 to Building 110. Calhoun let Richter know that this had slowed book writing to a trickle. In June of 1984, six months after he had at last settled into Building 110, Calhoun was told he needed to move yet again—this time to the Federal Building on Wisconsin Avenue in Bethesda. When he was first told he'd need to vacate Building 112, Calhoun had actually requested he be relocated to Bethesda (near the main campus of NIH), but it was another matter altogether to move the twenty-five miles from the NIH field station to the Federal Building after having lost months of time on the first relocation.

The same month that he was told he'd need to pack up and move again, Johns Hopkins University Press sent Calhoun a contract to sign for *The Rodent Key to Human Survival*. Calhoun forwarded the contract to Frederick Goodwin, the administrator at NIH who needed to approve such things and was informed that NIH would not sign the contract until after Calhoun moved into the Federal Building. Frustrated, Calhoun no doubt saw the second move and the contract delay as part of the "apparent fixed decision" to end his days at NIMH. He called Anders Richter at the Hopkins Press and explained that the second move would delay his book writing yet again.

While all this was going on, two pieces of good news came in. Calhoun was informed he was one of the recipients of the 1984 Presidential Design Awards for the human design ideas that emerged from his experimental work "exploring how the structure of the physical environment serves to influence the health and well-being of inhabitants." Then, in November of 1984, Calhoun learned Goodwin had changed his mind and would sign the book contract *as* Calhoun moved rather than waiting until the move was complete. That was all well and good, but it still took Calhoun and his team until April 20, 1985, before the move to the Federal Building was complete. By that point, more than two years had transpired since Richter had first contacted Calhoun. Between the two moves and NIH's bureaucratic intransience, Calhoun had written little more than the original book proposal. He called Richter to explain he'd need more time to complete the book than the contract dictated.

In addition to the roadblocks NIH was putting up, shortly after the move into the Federal Building, Calhoun, whose performance reviews had consistently been "outstanding" or "highly successful" was told that NIH "would not allow [him] to have a greater than 'satisfactory' performance until [he] completed [his] book." In response, Calhoun told administrators that he "could put up with [a review of] 'fully successful,' if [he] could just be left alone to complete [his] writing." But NIH didn't leave him alone. Instead, on December 5, 1985, though they continued to insist on external review at every stage of book production, NIH gutted the budget they had agreed to pay for seven external reviewers from $10,000 to $0. When Calhoun called Richter with an update, Richter told him of the "current uncertainties" already in place at the Press about publication of a book with so many logistic problems. Those uncertainties were now magnified by the lack of NIH support for their own review process. The contract was never officially canceled, but the book was now dead in its tracks.

Calhoun was close to his breaking point, but he made one last attempt to right things. In March 1986, he wrote Donald Ian MacDonald, the acting assistant secretary of Health and Human Services, whose purview included oversight of NIH. Calhoun requested a meeting to, he wrote, "discuss briefly with you in an unresolved situation which brings little credit to the National Institute of Mental Health or to your office." Calhoun continued, "As matters now stand it will be impossible to complete for publication the most important research of my professional career," referring to the NIH-caused debacles that appeared to end his chances of publishing a book that included results of his final experiments—and that would place those experiments within the context of more than four decades of work on population dynamics that had, he believed, profound implications for the problem of human overpopulation. "It grieves me particularly," Calhoun added, "not to have the opportunity to fulfill my stewardship . . . of public funds utilized in conducting . . . very large and very complex research projects."

Calhoun would remain grieved, for nothing came of his letter to MacDonald. By the fall of 1986, he'd had enough. On September 3, he wrote *New York Times* national correspondent John Herbers to tell him that the following day he was tendering his resignation, effective December 31: "My broader view of what humanity is all about," Calhoun told Herbers, "no longer conforms to the . . . policy of the National Institute of Mental Health." Calhoun packed up all his data files on the experiments he led since 1975, including the unpublished work on Universes 33 and 34 and his files and computer tapes on the electronic prosthesis gedankenexperiment and brought them to his home, where he had done some remodeling to create a large, comfortable workspace that he dubbed the Laboratory of Gnomonetics. A gnomon is an object whose shadow indicates the hour of the day (as on a sundial), but why Calhoun chose to call his home office the Laboratory of Gnomonetics, he didn't say—perhaps it was meant to convey that time was running out, and he would continue to work, this time from the shadows, to help solve the human population problem before it became unsolvable.[5]

After he retired, Calhoun would, on occasion, lecture about his work, as he did when he gave a talk at the Georgetown Family Center in Washington, DC. The talk was titled "A Hitchhiker's Guide to Calhoun's Rat Universes," a play on Douglas Adams's science fiction bestseller. In 1987, Calhoun submitted his last paper to *Transactions and Studies of the College of Physicians of Philadelphia*. That paper, "Population Extinction from Crowding: Induced Universal Altruism," included results from mouse Universe 33, but it was never published. In fact, none of the results from either mouse Universe 33 or rat Universes 34A and 34B were ever published. And the electronic prosthesis gedankenexperiment remained just that: a thought experiment that rarely left the confines of Calhoun's Laboratory of Gnomonetics.

17

GATHER ROUND, MY RATTIES

I'm truly sorry Man's dominion
Has broken Nature's social union,
An' justifies that ill opinion,
    Which makes thee startle,
At me, thy poor, earth-born companion,
    An' fellow-mortal!

. . . .

Still, thou art blest, compar'd wi' me!
The present only toucheth thee:
But Och! I backward cast my e'e,
    On prospects drear!
An' forward tho' I canna see,
    I guess an' fear!

ROBERT BURNS, "To a Mouse, on Turning Her Up in Her Nest, with the Plough," November 1785

The *New York Times* and the *Washington Post* both ran obituaries when seventy-eight-year-old John Calhoun died of a heart attack and stroke he suffered while on vacation in 1995. Calhoun, the *New York Times* wrote, "described a phenomenon in which some mice become 'beautiful ones,' maintaining their physical appearance, but doing little else, as the population swells," and added that Calhoun's "work was the inspiration for a children's book by Robert C. O'Brien, 'Mrs. Frisby and the Rats of NIMH.'" Both the *Times* and the *Washington Post* described the behavioral sink (though they didn't use the term), as well as a world full of Beautiful Ones. Although Calhoun came to think of his experimental work on population dynamics in rodents, particularly the work on rat cooperation and population growth, as a means of generating ideas to help avert the pitfalls of human overpopulation, the *Post* painted a different picture, writing that Calhoun saw his experiments as "grim forecasts for the future of the human race."

Major newspapers like the *New York Times* and the *Washington Post* were not the only ones paying tribute to Calhoun's work in 1995. Batman and Catwoman were as well. Two months after Calhoun died, in a Batman-laced issue of the *Catwoman* comic titled "The Secret of the Universe, Part Two," readers of DC Comics were reunited with archvillain Otis Flannegan, better known as Ratcatcher. Flannegan had indeed been a respectable ratcatcher in Gotham City before he turned to the dark side and obtained the unique ability to communicate with his former nemeses. "Gather round, my ratties," Ratcatcher tells his rodent troops, "and I'll tell you a tale to make your fur stand on end and your ratty blood run cold." That tale is where Calhoun and his experiments enter the picture. "Some behaviorists made a nice, big cage with pipes and tunnels and gnawing blocks and everything a happy rat could want," Ratcatcher explains, swapping rats for the mice that were in Calhoun's Universe 33. "Food and drink were free and regular and of the highest standard, my sources say. They called it universe 133." All went well in universe 133, Ratcatcher tells his enthralled followers: "They played, they ate, they bred. No problem. The food was unlimited, see. But the living space wasn't. Numbers grew but there was nowhere to go. Discipline broke down. Grooming ceased. Gangs of adolescent males roamed the universe bullying and killing." Ratcatcher gives a fairly accurate description of Calhoun's behavioral sink, and as he does, the rats in the audience take on much more menacing looks as they learn of the horrors within universe 133. "Rat turned against rat in an orgy of violence," Ratcatcher tells them. "Cannibalism was rife. The colony died en masse and in the final hours the most hideous act of all . . . when the mothers devoured their own babies." Ratcatcher would have no part of it: "Sickening, isn't it? And so like humans to drive an innocent species to such unparalleled barbarism," he tells his rodent followers. "I look at Gotham my friends and what do I see? Gangs roam the streets attacking at will. Stress-related illness! Crime! Murder! Their society is collapsing around them just like universe 133."[1]

Thanks not just to Batman and Catwoman, but to periodic popular science articles, podcasts, and more, Calhoun's work lives on in our collective consciousness. *Cabinet* magazine's 2011 online article, "The Behavioral Sink: The Mouse Universes of John B. Calhoun," tells readers of mouse life in the utopia-turned-hell world of Universe 25, as well as the tale of rats in the Casey Barn. "Calhoun's research remains a touchstone for a particular kind of pessimistic worldview," Will Wiles tells his readers. "In the way that writers like Wolfe and the historian Lewis Mumford deployed reference to it, it can be seen as bleakly reactionary, a warning against cosmopolitanism or welfare dependence, which might sap the spirit and put us on the skids to the behavioral sink." The following year, film clips of Calhoun, along with his mice and rats, appeared in *Critical Mass*, an environmental documentary on the population explosion in humans.

In 2015, *Gizmodo* ran a story, titled "How Mice Turned Their Private Paradise into a Terrifying Dystopia," that regaled readers with tales of the Beautiful Ones in Universe 25 who were "spared from violence and death, but had completely lost touch with social behaviors, including having sex or caring for their young." After describing the story of the collapse of mouse society, Esther Inglis-Arkell ends her article asking readers, "Can we escape Universe 25's fate?" The following year, *Atlas Obscura* published an article, "The Doomed Mouse Utopia That Inspired the 'Rats of NIMH': Dr. John Bumpass Calhoun Spent the '60s and '70s Playing God to Thousands of Rodents." Here, readers learned not just of Universe 25 and the possible link between Calhoun's work and *Mrs. Frisby and the Rats of NIMH*, but that Calhoun "extrapolated . . . to human concerns, becoming an early supporter of environmental design and H.G. Wells's hypothetical 'World Brain,' an international information network that was a clear precursor to the internet."

The *Washington Post*, which had been so integral in spreading word of Calhoun's work in the 1960s and 1970s, returned to it once again in 2017, in Fredrick Kunkle's article, "The Researcher Who Loved Rats and Fueled Our Doomsday Fears." To prepare a new generation of *Post* readers to learn of Calhoun and his work, Kunkle wrote about a 1971 profile the *Post* had published on Calhoun, in which Calhoun had said, "Rats and mice, of course, are not perfect models of humans, but the disaster they represent is so compelling that the world cannot wait for proof of every step in the equation."

Washington Post readers in 2017 learned of high-rise rodent apartment complexes, pied pipers, Beautiful Ones, and more. And, for once—despite the title of the article—at the end Calhoun is not portrayed as a prophet of doom but an optimist. Kunkle explained why after first acquainting readers with those creative rats who had come up with the rodent equivalent of the wheel in the Towson enclosure: "Just as Calhoun had altered the animals' behavior by tinkering with the colony's physical design he believed that humans could counter the effects of overcrowding by modifying their environment. Through technology and culture, people could enlarge the 'conceptual space' that allows them to live in peace among a multitude."[2]

That is indeed what Calhoun believed. Whether technology can increase our conceptual space so that we live in peace is an open question, but regardless of the ultimate answer, Calhoun's work on rats and mice shows us, in its rawest form, the existential dangers of overpopulation.

EPILOGUE

In his 2017 story, Kunkle told *Washington Post* readers that Calhoun was a man "who achieved the recognition attained by only a handful of other social scientists, such as Pavlov and Skinner." Certainly, in the 1960s and early 1970s, Calhoun's work generated tremendous attention, largely positive, but some critical. But the lasting impact of his work is nowhere near that of Pavlov's work or Skinner's work, which have shaped and continue to shape research programs around the world and are discussed, at length, in nearly every psychology textbook, as well as many an animal behavior textbook (including my own).

It's true that, in its heyday, Calhoun's work on rats and mice could also be found in animal behavior and psychology textbooks. But a review of textbooks in those fields today shows virtually no mention of Calhoun's studies. And to find a discussion of Calhoun's work in academic journals today is no easy matter either. While historians Jon Adams and Edmund Ramsden have written a series of excellent papers that look at Calhoun's work, those papers are published in journals such as *Comparative Studies in Society and History* and *Journal of Social History*—journals that researchers studying evolution, animal behavior, population dynamics, and psychology would rarely, if ever, read. On occasion, a PhD dissertation will still cite one or more of Calhoun's studies, but these tend to be citations to show that the author has done their history homework rather than citations suggesting that Calhoun's work helped shaped the ideas in the dissertation. Web of Science, the leading citation index for scientific papers, shows that almost none of Calhoun's papers are cited anymore. The one exception is Calhoun's 1973 paper "Death Squared," which is still cited on occasion, although virtually never in a major animal behavior or psychology journal.[1]

Why has Calhoun's work not fared well in the world of academia? For one thing, after he published "Death Squared" in the *Proceedings of the Royal Society of Medicine*, Calhoun rarely published his results in mainstream science journals, and, indeed, many of his studies, including those on Universes 33 and 34, were never published at all. Word of those studies would only have spread through people who heard Calhoun lecture about them or read a mention of them in a review paper Calhoun wrote in his later years.

Calhoun's sometimes glib use of anthropomorphic terminology has also hurt his standing in the world of science. Today, the sort of anthropomorphic language—Beautiful Ones, universal autism, pied pipers, somnambulists, and more—that Calhoun used, even in his most technical papers, is not just frowned upon but thought of as unprofessional, as well as dangerous. It's difficult to imagine an editor of a major journal in animal behavior, evolution, or psychology allowing an author to describe extremely aggressive individuals as "berserk," as Calhoun did. Anthropomorphism was frowned on at the time Calhoun was doing his experiments as well, but there was much more leeway then. In a similar vein, while most scientists both today and in Calhoun's day would agree that behavioral and evolutionary work in nonhumans can inform our understanding of ourselves, the idea that a handful of studies in a small number of species could or should impact policy decisions is approached with much greater trepidation. But, both in his papers and in his lectures, Calhoun would often slip into language that, at the very least, made it appear as if his policy recommendations did indeed stem largely from his work on two species of rodents.

In the field of animal behavior, and to a lesser extent in psychology, another reason Calhoun's work has fallen off the map is that there has been a significant shift in perspective toward detailed cost-benefit analyses in the study of nonhuman behavior. Calhoun considered both the costs and benefits of social behavior in his universes, but he wasn't all that concerned with directly measuring those costs and benefits or casting his work in an explicit, cost-benefit framework. For example, as opposed to the approach Calhoun took, a study of aggression today might measure the size of the combatants, the amount of energy expended on different types of fighting behavior, the risk associated with different levels of fighting, the benefits linked to winning a fight, and more. Work on population dynamics and behavior has also changed radically since Calhoun undertook his experiments. To see how, recall the experiment on social networks and mice touched on in the preface.

In that ongoing study twenty-five kilometers northwest of Zurich, Switzerland, Barbara König and her colleagues partitioned the floor of a barn into four sections, each with ten nests, three water fountains, and ten feeding stations. The barn today is home to about four hundred house mice, all descended from four males and eight females that König and her colleague, Andrea Weidt, caught (in 2002) within three miles of the barn. Each mouse in the barn now has a very small chip, which weighs slightly more than half an ounce, implanted under its skin. The chip emits a signal that is detected whenever a mouse enters or exits its nest and when it is near a water fountain. This produces a nonstop stream of data on who is in a nest, when they enter and exit, and who they are interacting with outside the nest at drinking stations. What's more, digital scales with infrared motion detectors are under some of the drinking stations so that König knows the weight of the mice (a good indicator of health) and how their weight changes over time. At the end of each day, all this information is automatically transmitted to a server at the University of Zurich, at which point it is run through a computer program that translates the raw data into a form that allows the researchers to construct mouse social networks within the population in the barn.

All that technology is eye popping, but it is the data that tells us about behavior and population dynamics. König and her team have used that data to generate detailed social networks in the barn mice: networks revolving around family dynamics, the spread of disease, and more. But what makes this work so different from the plethora of social network studies that fill the pages of today's animal behavior journals is what König and her team learned about social networks and population dynamics when a cat snuck into their barn on the weekend of January 19, 2019.

Population biologists and animal behaviorists both have long been interested in how populations respond to catastrophic events, but relatively little is known about such responses. The cat that somehow snuck into König's barn wreaked havoc and, in so doing, provided the "cat" in the catastrophe that would lead König and her team to employ social network analysis to see how the barn population coped.

Of the 478 chip-tagged mice there in the barn the week before the cat struck, eight-five died. In addition, there were thirty-two untagged victims, and another hundred or so mice were missing. House mice reproduce quickly, and the population began to recover in a matter of months. But, from the day after the cat struck, König and her team gathered data on how this large-scale predation event affected the social network of those mice that had survived. Of the seventeen mouse networks in the barn before disaster struck, fourteen remained, but each had lost members, with mortality ranging from 12 percent to a devastating 88 percent. König knew from other work that house mouse networks were complex, and the researchers were about to find out just how complex.

After the cat struck, mice who were connected to relatively few others in their network before the "cat-astrophe" formed many new, but weak, bonds with others, perhaps seeking safety in numbers.

Mice who were well connected before the predation event had fewer associates after the cat attacked, but the bonds they formed were stronger than before: these mice became more "socially insular" but compensated by being involved in more prosocial interactions with their nest-box mates.

König and colleague's work illustrates the sort of complex social network cost-benefit analysis of population dynamics and behavior that permeates the literature today. A few years back, I asked König if she knew of Calhoun's work on population dynamics in rodents. She paused, unsure, until I mentioned mouse utopias, at which point she had a vague recollection, but no more, of coming across Calhoun's studies at some time or another.

Others have more vivid recollections of Calhoun, his work, and its long-term impact (or lack thereof). When Steve Suomi, who moved into Calhoun's space at NIMH, looks back on the body of Calhoun's work, he's most impressed by one thing: "He anticipated," Suomi says today, "the cross-disciplinary integration necessary to really study complex developmental problems." Powerful praise, indeed, for Calhoun the thinker. Neil Greenberg, an animal behaviorist who, as a postdoc, overlapped briefly with Calhoun at NIMH in the 1970s, casts Calhoun's work as the sort of material that gets undergraduate students excited about the study of behavior. "I ended up using some of his research as gee-whiz kind of stuff for teaching," he says today.[2]

Perhaps more than anything else, the reason references to Calhoun's work on his rodent universes have plummeted in the scientific literature is because no evidence for the behavioral sinks, Beautiful Ones, or other observations that Calhoun detailed in his rat and mouse universes has been found in wild populations of animals—rat, mouse, or otherwise. Part of that is, no doubt, because for the last few decades, no one has been explicitly looking for these things when working in the field. But in the 1970s and early 1980s, most animal behaviorists in both biology and psychology departments would have been familiar with Calhoun's work, yet field studies then (and later) on population dynamics were not finding anything that resembled behavioral sinks, Beautiful Ones, and so on.

Unlike its fate in the academy and its technical journals, between popular science articles, podcasts, and the occasional mention in newspapers, Calhoun's work has remained a subject of fascination in popular culture (and perhaps its success in that culture is due to the same reason that it hasn't fared well over the long run in scientific circles). After all, how could the public not be endlessly fascinated with the United States Senate discussing rodent utopias-turned-dystopias full of Beautiful Ones scampering about rodent apartment complexes and pied piper mice following around a researcher who told a collection of the best scientists in the world that his work may "sound like rantings of a mad egghead locked up in his ivory tower," and who wanted to write a book called *The Rodent Key to Human Survival*?

ACKNOWLEDGMENTS

The John Calhoun Archive at the National Library of Medicine in Bethesda, Maryland, was invaluable in writing this book. My faithful research assistant, Robin Zecca, spent endless hours in that archive, scanning thousands of pages of Calhoun material and posting that material for me to dig into.

This book would not be the same without interviews conducted with Catherine Calhoun, Cheshire Calhoun, Jonathan Freedman, Neil Greenberg, Kathy Kerr, Vernon Reynolds, Norman Slade, Steve Suomi, and Stuart Umpleby. I thank them all for taking time from their busy schedules to chat with me. Catherine Calhoun was also kind enough to send me numerous documents from her own archives of material on her father and his work.

My sincere thanks to the finest editor I know, Joe Calamia, whose input at every stage in the book-writing process was invaluable. Joe is a true pleasure to work with. The entire team at the University of Chicago Press helped make this book what it is and I am in debt to Nick Lilly, marketing manager; Matt Lang, editorial associate; Lindsy Rice, manuscript editor; and Rae Ganci Hammers, senior designer, for all their fine work. Special thanks are due to Dana Dugatkin, Aaron Dugatkin, Henry Bloom, and Michael Sims for reading various sections of the book as it was being written, and for providing important and insightful comments.

Last, but not least, I would like to thank Julia Rene I, Lena, Leisha, 2R, Oliver, Blue, and the Hip Cheep for all they have taught me about the complex dynamics of rodent society.

NOTES

PREFACE

1. "Calhoun's . . . Norway rat colonies": E. O. Wilson, *Sociobiology: The New Synthesis* (Cambridge, MA: Harvard University Press, 1975), 84; "effects [that] were observed": Wilson, *Sociobiology*, 255; "Rats and mice, of course": T. Huth, "Of Mice and Men," *Washington Post*, March 29, 1971. That interview was entered into the Congressional Record, Proceedings and Debates of the 92nd Congress, First Session, April 1, 1971, S4398–S4399. An index of the National Library of Medicine's John B. Calhoun Papers can be found here: https://findingaids.nlm.nih.gov/repositories/ammp/resources/calhoun586.

   The textbooks referred to are L. A. Dugatkin, *Principles of Animal Behavior* (New York: W. W. Norton, 2004); L. A. Dugatkin, *Principles of Animal Behavior*, 5th ed. (Chicago: University of Chicago Press, forthcoming); and C. T. Bergstrom and L. A. Dugatkin, *Evolution*, 3rd ed. (New York: W. W. Norton, 2023).

INTRODUCTION

1. "I shall largely speak of mice": J. Calhoun, "Death Squared: The Explosive Growth and Demise of a Mouse Population," *Proceedings of the Royal Society of Medicine* 66 (January 1973): 80–88, 80; "a Utopian environment": Calhoun, "Death Squared," 81; "four four-unit": Calhoun, "Death Squared," 82; "pathological togetherness": J. Calhoun, "Population Density and Social Pathology," *Scientific American* 206, no. 2 (1962): 139–48, 139; "Normal social organization" and "capable only of the most simple": Calhoun, "Death Squared," 86; T. Malthus, *An Essay on the Principle of Population* (London: J. Johnstone, 1978); P. Ehrlich, *The Population Bomb* (New York: Ballantine Books, 1968).
2. S. Alsop, "Dr. Calhoun's Horrible Mousery," *Newsweek*, August 17, 1970, 96; "even if some way can be found": O. Friedrich, "Population Explosion: Is Man Really

Doomed?," *Time*, September 13, 1971; Congressional Record, Proceedings and Debates of the 92nd Congress, First Session, April 1, 1971, S4398–S4405; R. Hock, *Forty Studies That Changed Psychology*, 8th ed. (New York: Pearson, 2020); T. Wolfe, *The Pump House Gang* (New York: Farrar, Straus & Giroux, 1968); A. Grant, J. Balent, and B. Smith, "The Secret of the Universe, Part Two," *Catwoman* 2, no. 26, November 1995; R. O'Brien, *Mrs. Frisby and the Rats of NIMH* (New York: Atheneum Books, 1971); "We are now at the critical transition": J. B. Calhoun, "Promotion of Man," in *Global Systems Dynamics: International Symposium, Charlottesville, Va., 1969*, ed. E. O. Attinger (Basel, Switzerland: S. Karger, 1969), 36–58 and 63–65, 63; "a rough calculation indicates": Calhoun, "Promotion of Man," 62.

CHAPTER 1

1. "tended to produce teachers": J. Calhoun, "What Sort of Box?," *Man-Environment Systems* 3 (1973): 3–30, 12; "Tennessee country boy": J. Calhoun, "Two Tennessee Country Boys: A Eulogy for Murry Bowen," 1990, Bowen Center for the Study of Family, provided by Kathy Kerr; "Naturally, the home situation": Boyhood Recollections of John Bumpass Calhoun by His Mother, Fern M. Calhoun, n.d., p. 1, box 135, folder 12, John B. Calhoun Papers, National Library of Medicine (henceforth NLM); "sing and dance": Boyhood Recollections, n.d., NLM, 1–2; "first memory": J. Calhoun, "Looking Backward from 'The Beautiful Ones,'" in *Discovery Processes in Modern Biology: People and Processes in Biological Discovery*, ed. W. Klemm (New York: Krieger, 1977), 31; "physically isolated": Calhoun, "Looking Backward," 31; "chewing on apple seeds": Calhoun, "Looking Backward," 32; "Hunting—we received shotguns" and information about Dinah Shore: Concerning the Development of a Point of View: An Autobiography by John B. Calhoun, 1962, box 23, folder 57, John B. Calhoun Papers, NLM; "watching doves": Calhoun, "Looking Backward," 31–32.
2. "being small": Boyhood Recollections, n.d., NLM, 4; "On the spot": Calhoun, "Looking Backward," 33–34; told students of H. V. Wilson's work: Tribute dinner for John Calhoun, 1995, provided by Cheshire Calhoun; H. V. Wilson, "On Some Phenomena of Coalescence and Regeneration in Sponges," *Journal of Experimental Zoology* 5 (1907): 245–58. For more, see A. Ereskovsky, I. E. Borisenko, F. V. Bolshakov, and A. I. Lavrov, "Whole-Body Regeneration in Sponges: Diversity, Fine Mechanisms, and Future Prospects," *Genes* 12, no. 4 (2021): 506, https://doi.org/10.3390/genes12040506; "From Kepner came," "sharpening one's," and "a 'gentleman's' school": Calhoun, "Looking Backward," 34; A. Kinsey, *The Gall Wasp Genus Cynips: A Study in the Origin of Species*, Indiana University Studies vol. XVI (Bloomington: University of Indiana, 1930).
3. J. B. Calhoun, "Swift Banding at Nashville and Clarksville," *Migrant* 9 (1938): 77–81, 78; J. B. Calhoun and J. C. Dickinson Jr., "Migratory Movements of Chimney Swifts, *Chaetura pelagica* (Linnaeus) Trapped at Charlottesville, Virginia,"

Bird-Banding 13, no. 2 (1942): 56–59; J. B. Calhoun, "Notes on the Summer Birds of Hardeman and McNairy Counties," *Tennessee Academy of Science* 16 (1941): 293–309. Also see B. Coffey, "Swift Banding in the South," *Migrant* 9 (1938): 82–84; and H. Peters, "Chimney Swift Banding in Alabama during the Fall of 1936," *Bird-Banding* 8, no. 1 (1937): 16–24.

CHAPTER 2

1. "general economy of the household of nature" and "scientific natural history": A. S. Pearse, *Animal Ecology*, 2nd ed. (New York: McGraw-Hill, 1939), 1–4; "the sociology of organisms": V. Shelford, *Laboratory and Field Ecology* (Baltimore: Williams & Wilkins Co., 1929), 608; "an animal consists of matter" and "most scientists": Pearse, *Animal Ecology*, 14. For more on Haeckel and ecology, see F. N. Egerton, "Ernst Haeckel's Ecology," *Bulletin of the Ecological Society of America* 94, no. 3 (2013): 222–44. For more on the British Ecological Society, see E. Salisbury, "The Origin and Early Years of the British Ecological Society," *Journal of Animal Ecology* 33 (1964): 13–18.
2. R. Pearl, "The Movements and Reactions of Freshwater Planarians" (PhD diss., University of Michigan, 1902). Pearl's mentor in biometry, Karl Pearson, was a student of Sir Francis Galton, Charles Darwin's cousin. Pearl's work on the logistic equation built on prior work by Pierre-François Verhulst and Alfred J. Lotka. R. Pearl and L. Reed, "On the Rate of Growth of the Population of the United States since 1790 and Its Mathematical Representation," *Proceedings of the National Academy of Sciences of the United States of America* 6, no. 6 (1920): 275–88; R. Pearl, *The Biology of Population Growth* (Baltimore: Williams & Wilkins Co., 1925). For more on Pearl, see S. Kingsland, *Modeling Nature: Episodes in the History of Population Ecology* (Chicago: University of Chicago Press, 1988); S. Kingsland, *The Evolution of American Ecology, 1890–2000* (Baltimore: Johns Hopkins University Press, 2005); and E. Ramsden, "Carving up Population Science: Eugenics, Demography and the Controversy over the 'Biological Law' of Population Growth," *Social Studies of Science* 32, no. 5/6 (2002): 857–99.
3. "willingness to be influenced": J. Calhoun, "Looking Backward from 'The Beautiful Ones,'" in *Discovery Processes in Modern Biology: People and Processes in Biological Discovery*, ed. W. Klemm (New York: Krieger, 1977), 35–36; O. Park, *Sherlock Holmes, Esq., and John H. Watson, M.D.: An Encyclopaedia of Their Affairs* (Evanston, IL: Northwestern University Press, 1962); "It was common": M. Engelmann, "Resolution of Respect: Orlando Park, 1901–1969," *Bulletin of the Ecological Society of America* 51, no. 1 (1970): 16–20. A list of many of Orlando Park's papers can be found in W. C. Allee, O. Park, A. E. Emerson, T. Park, and K. P. Schmidt, *Principles of Animal Ecology* (Philadelphia: Saunders Company, 1949), 777–78; P. Rau, "Population Studies in Colonies of *Polistes* Wasps; with Remarks on the Castes," *Ecology* 20, no. 3 (1939): 439–42; W. F. Blair, "Notes on Home Ranges and

Populations of the Short-Tailed Shrew," *Ecology* 21, no. 2 (1940): 284–88; W. J. Hamilton Jr., "Winter Reduction of Small Mammal Populations and Its Probable Significance," *American Naturalist* 76, no. 763 (1942): 216–18; "Symposium: Population Problems in Protozoa," *American Naturalist* 75, no. 760 (1941).

4. "long meandering lines" and "I can remember when the thought struck me": Calhoun, "Looking Backward," 36. The paper that Calhoun read on grid sampling was B. P. Bole Jr., *The Quadrat Method of Studying Small Mammal Populations* (Cleveland: Cleveland Museum of Natural History, Scientific Publications, 1939), 15–177; J. B. Calhoun, "Distribution and Food Habits of the Mammals in the Vicinity of Reelfoot Lake Biological Station," *Tennessee Academy of Science* 17 (1941): 177–85 and 207–25; "I was much interested": Lotka to Calhoun, April 28, 1941, box 1, folder 19, John B. Calhoun Papers, NLM; "At this time of intellectual searching," "by the end," and "the possibility of fusing": Calhoun, "Looking Backward," 36; "We view the population": Allee et al., *Principles of Animal Ecology*, 6; "I fell under the influence": Vita and Theoretical Framework Prepared for Dr. Leonard Carmichael, 1954, box 23, folder 18, John B. Calhoun Papers, NLM.
5. "the border-line field": W. C. Allee, *Animal Aggregations: A Study in General Sociology* (Chicago: University of Chicago Press, 1931), 9; "Even though one should admit": W. C. Allee, "Cooperation among Animals," *University of Chicago Magazine*, June 1928, 419; W. C. Allee, *The Social Life of Animals* (New York: W. W. Norton, 1938), republished as *Cooperation among Animals with Human Implications* (New York: Henry Schuman, 1951); Allee's 1928 lectures at Northwestern: Harris Foundation lectures, 1937–1938, box 6, folder 6, Allee archives, University of Chicago Library; "under the tutelage": J. Calhoun, "What Sort of Box?," *Man-Environment Systems* 3 (1973): 4. For more on Allee, see K. P. Schmidt, "Warder Clyde Allee," *Proceedings of the National Academy of Sciences, Biographical Memoirs* 30 (1957): 3–40; L. A. Dugatkin, *The Altruism Equation: Seven Scientists Search for the Origins of Goodness* (Princeton, NJ: Princeton University Press, 2006); G. Mitman, "From the Population to Society: The Cooperative Metaphors of W.C. Allee and A.E. Emerson," *Journal of the History of Biology* 21, no. 2 (1988): 173–94; and G. Mitman, *The State of Nature: Ecology, Community, and American Social Thought, 1900–1950* (Chicago: University of Chicago Press, 1992).
6. "looking at social relations" and "World War II necessitated": Calhoun, "Looking Backward," 36; J. B. Calhoun, "An Analysis of the Locomotor Activity Rhythms of Two Ecologically Equivalent Rodents, *Microtus ochrogaster* (Wagner) and *Sigmodon hispidus hispidus* Say and Ord" (PhD diss., Northwestern University, 1943). Calhoun published the results from his dissertation in J. B. Calhoun, "Diel Activity Rhythms of the Rodents, *Microtus ochrogaster* and *Sigmodon hispidus hispidus*," *Ecology* 26, no. 3 (1945): 251–73; "In the seven days from July 31": Calhoun, "Diel Activity Rhythms," 263; "to try to understand": J. Calhoun, "Jacob's Ladder," 1992, box 132, folder 40, John B. Calhoun Papers, NLM.

CHAPTER 3

1. "Genetic adaptation intrigued me": J. Calhoun, "Looking Backward from 'The Beautiful Ones,'" in *Discovery Processes in Modern Biology: People and Processes in Biological Discovery*, ed. W. Klemm (New York: Krieger, 1977), 37; "For me, teaching the mass-standardized," "left [him] focused," "Scientifically this wasn't," and "Watching the rats": Calhoun, "Looking Backward," 38; J. B. Calhoun, "Population Cycles and Gene Frequency Fluctuations in Foxes of the Genus *Vulpes*, in Canada," *Canadian Journal of Research* 28 (1950): 45–57; J. B. Calhoun, "Utilization of Artificial Nesting Substrate by Doves and Robins," *Journal of Wildlife Management* 12, no. 2 (1948): 136–142; F. Morton, *A Social Study of Ward 5 and 10 in Baltimore, Maryland* (Baltimore: Baltimore Council of Social Agencies, 1937). For more on housing in Baltimore at this time see B. Leclair-Paquet, "The 'Baltimore Plan': Case-Study from the Prehistory of Urban Rehabilitation," *Urban History* 44, no. 3 (2017): 516–43; M. Hoffman, "Changes in Baltimore's Housing, 1940 to 1950, with Special Reference to the Role of Government" (MA thesis, American University, 1953); and "A Law Spelling Out the 'Baltimore Plan,'" *Baltimore Sun*, February 7, 1951.
2. "Soon I found myself": Calhoun, "Looking Backward," 38.
3. "Mr. Richter, I do not": C. P. Richter, "It's a Long, Long Way to Tipperary, the Land of My Genes," in *Leaders in the Study of Animal Behavior: Autobiographical Perspectives*, ed. D. A. Dewsbury (Lewisberg, PA: Bucknell University Press, 1985), 356–86, 367; "This course, though short": Richter, "It's a Long, Long Way," 369; 200-million-dollar estimate: "Rats are Costly Pests and a Danger to Health," *Science News*, June 13, 1942, 274; "I shall never forget": C. P. Richter, "Experiences of a Reluctant Rat-Catcher: The Common Norway Rat—Friend or Enemy?," *Proceedings of the American Philosophical Society* 112, no. 6 (1968): 403–15, 406. For more on rodenticide work during World War II, see C. Keiner, "Wartime Rat Control, Rodent Ecology, and the Rise and Fall of Chemical Rodenticides," *Endeavour* 29, no. 3 (2005): 119–25; and J. C. Adams and E. Ramsden, "Rat Cities and Beehive Worlds: Density and Design in the Modern City," *Comparative Studies in Society and History* 53, no. 4 (2011): 722–56. For more on Richter and rat foraging, see C. P. Richter, "Animal Behavior and Internal Drives," *Quarterly Review of Biology* 2, no. 3 (1927): 307–43; and C. P. Richter and B. Barelare Jr., "Further Observation on the Carbohydrate, Fat, and Protein Appetite of Vitamin B Deficient Rats," *American Journal of Physiology* 127 (1939): 199–210. For more on Richter in general and with respect to ANTU, see C. P. Richter and K. H. Clisby, "Toxic Effects of the Bitter-Tasting Phenylthiocarbamide," *Archives of Pathology* 33, no. 1 (1942): 46–57; C. P. Richter, "The Development and Use of Alpha-Naphthyl Thiourea (ANTU) as a Rat Poison," *Journal of the American Medical Association* 129 (1945): 927; J. Schulkin, *Curt Richter: A Life in the Laboratory* (Baltimore: Johns Hopkins

University Press, 2005); J. Schulkin, "Curt Richter: Psychobiology and the Concept of Instinct," *History of Psychology* 10, no. 4 (2007): 325–43.

4. "There are no strings attached": Keiner, "Wartime Rat Control," 121–22; "fifteen young, enthusiastic air-raid wardens": Richter, "Experiences of a Reluctant Rat-Catcher," 407; "Always an exciting occasion": Richter, "Experiences of a Reluctant Rat-Catcher," 408; J. T. Emlen Jr., *Adventure Is Where You Find It: Recollections of a Twentieth Century Naturalist* (self-pub., 1996); C. P. Richter and J. T. Emlen Jr., "A Modified Rabbit Box Trap for Use in Catching Live Wild Rats for Laboratory and Field Studies," *Public Health Reports* 60, no. 44 (1945): 1303–8. For more on Richter, Davis, Calhoun, and the Rodent Ecology Project, see C. P. Richter, "Experiences of a Reluctant Rat-Catcher"; J. T. Emlen Jr., "Baltimore's Community Rat Control Program," *American Journal of Public Health* 37, no. 6 (1947): 721–27; E. Ramsden, "From Rodent Utopia to Urban Hell: Population, Pathology, and the Crowded Rats of NIMH," *Isis* 102, no. 4 (2011): 659–88; E. Ramsden and J. C. Adams, "Escaping the Laboratory: The Rodent Experiments of John B. Calhoun and Their Cultural Influence," *Journal of Social History* 42, no. 3 (2009): 761–92; E. Ramsden, "Rats, Stress and the Built Environment," *History of the Human Sciences* 25, no. 5 (2012): 123–47; and E. Ramsden, "Travelling Facts about Crowded Rats: Rodent Experimentation and the Human Sciences," in *How Well Do Facts Travel?*, eds. M. Morgan and P. Howlett (Cambridge: Cambridge University Press, 2011), 223–51.
5. "effectively an island": J. J. Christian and D. E. Davis, "The Relationship between Adrenal Weight and Population Status of Urban Norway Rats," *Journal of Mammalogy* 37, no. 4 (1956): 475–86, 476; "What was a block-box," "yards usually contained," "climbing high board fences," and "the private lives of rats": J. Calhoun, "What Sort of Box?," *Man-Environment Systems* 3 (1973): 5.

CHAPTER 4

1. The Davis pilot experiment is discussed in D. E. Davis, "Early Behavioral Research on Populations," *American Zoologist* 27, no. 3 (1987): 825–37; "override . . . rat clannishness": J. Calhoun, "What Sort of Box?," *Man-Environment Systems* 3 (1973): 6; "They tried to help me" and "all hell broke loose": Calhoun, "What Sort of Box?," 6.
2. "survival required": Calhoun, "What Sort of Box?," 6; "Psychological turmoil": D. E. Davis, "The Role of Intraspecific Competition in Game Management," in *Transactions of the Fourteenth North American Wildlife Conference* (Washington, DC: Wildlife Management Institute, 1949), 225–31, 231; "Rats are as bad as," "the troubles which," and "In a long-established rat community": "Displaced Rats," *Time*, June 14, 1948. John Christian, another member of the Rodent Ecology Project, found that as population size and aggression increased, the adrenal gland, responsible for the production of stress hormones like cortisol, increased as

well, suggesting an underlying hormonal link to stress as a break on population growth. For more on Christian's work on the adrenal gland and stress, see J. Christian, "The Adreno-Pituitary System and Population Cycles in Mammals," *Journal of Mammalogy* 31, no. 3 (1950): 247–59; J. Christian and D. E. Davis, "The Relationship between Adrenal Weight and Population Status of Urban Norway Rats," *Journal of Mammalogy* 37, no. 4 (1956): 475–86; D. E. Davis and J. Christian, "Changes in Norway Rat Populations Induced by Introduction of Rats," *Journal of Wildlife Management* 20, no. 4 (1956): 378–83; and J. Christian, "Phenomena Associated with Population Density," *Proceedings of the National Academy of Sciences of the United States of America* 47, no. 4 (1961): 428–49.

3. "a habitat that simulated": J. Calhoun, "Looking Backward from 'The Beautiful Ones,'" in *Discovery Processes in Modern Biology: People and Processes in Biological Discovery*, ed. W. Klemm (New York: Krieger, 1977), 39; "Not knowing really what I meant": Calhoun, "What Sort of Box?," 7; "enable the formation and delimitation": J. Calhoun, *The Ecology and Sociology of the Norway Rat* (Bethesda, MD: US Department of Health, Education, and Welfare, 1962), 5; "the rat is a social animal": Calhoun, *Ecology and Sociology*, 245.
4. "actually a conservative" and "50,000 descendants": Calhoun, *Ecology and Sociology*, 245.
5. "The best estimate," and "[What] was the cause of this 25-fold decrease": Calhoun, *Ecology and Sociology*, 245; "develop a social organization": Calhoun, *Ecology and Sociology*, 136; "All the evidence is": Calhoun, *Ecology and Sociology*, 244; "A stable group is one in which": J. B. Calhoun, "The Social Aspects of Population Dynamics," *Journal of Mammalogy* 33, no. 2 (1952): 139–59, 141; "Most sex for them was homo[sexual]": Calhoun, "What Sort of Box?," 8; J. Calhoun, "A Method for Self-Control of Population Growth among Mammals Living in the Wild," *Science* 109 (1949): 333–35; "floated many trial balloons": Calhoun, "Looking Backward," 40; "could never fulfill": Calhoun, "Looking Backward," 42.
6. "Human populations are approaching": J. Calhoun, "The Study of Wild Animals under Controlled Conditions," *Annals of the New York Academy of Sciences* 51, no. 6 (1950): 1113–22, 1122; W. Vogt, *Road to Survival* (Boston: Little, Brown & Co., 1948); F. Osborn, *Our Plundered Planet* (Boston: Little, Brown & Co., 1948); "There are certain methodological problems": Marshak to Calhoun, March 8, 1949, box 1, folder 26, John B. Calhoun Papers, NLM; "the techniques of experimentation": Calhoun to Marshak, March 26, 1949, box 1, folder 26, John B. Calhoun Papers, NLM; "Quite recently": Marshak to Calhoun, June 25, 1949, box 1, folder 26, John B. Calhoun Papers, NLM.

CHAPTER 5

1. "the relationship between heredity" and "The study of such immense": Minutes of the Conference on Genetics and Social Behavior, 1946, box 6, folder 22,

John B. Calhoun Papers, NLM; "They [NIMH] have become quite interested": Calhoun to Marshak, March 26, 1949, box 1, folder 26, John B. Calhoun Papers, NLM; At the University of Michigan, Calhoun visited the lab of Lee Dice. For more on the Jackson Memorial Laboratory, see D. Dewsbury, "A History of the Behavior Program at the Jackson Laboratory: An Overview," *Journal of Comparative Psychology* 126, no. 1 (2012): 31–44; and https://www.jax.org/about-us/history #at-jax. For more on the Committee for the Study of Animal Societies under Natural Conditions, see "Forty-Sixth Annual Meeting of the American Society of Zoologists, New York City, December 28, 29, 30, 1949," *Anatomical Record* 105 (1949): 451–633.

2. For more on "Basic Research into Problems of Behavior and its Application to Problems of Human Welfare," see Basic Research into Problems of Behavior and Its Application to Problems of Human Welfare, 1949, box 23, folder 7, John B. Calhoun Papers, NLM. For more on "Effects of Early Experience on Mental Health," see *Minutes of the Conference on the Effects of Early Experience on Mental Health, September 6–9, 1951* (Bar Harbor, ME: Roscoe B. Jackson Memorial Laboratory, 1951); and Minutes of the Conference on the Effects of Early Experience on Mental Health, 1951, box 6, folder 28, John B. Calhoun Papers, NLM; B. F. Skinner, *The Behavior of Organisms* (New York: Appleton-Century-Crofts, 1938); B. F. Skinner, *Walden Two* (Indianapolis, IN: Hackett Publishing, 1948). Before Skinner, at the turn of the twentieth century, Edward Thorndike was looking at instrumental learning by testing how quickly cats could learn to escape from his "puzzle boxes." The cats tried all sorts of things to get out of their confined spaces. Some behaviors, by chance, led to a successful escape from the box. Thorndike hypothesized that cats began to pair certain behaviors that they undertook in the box with a positive effect—escape—and they were then more likely to use such behaviors when confined in the puzzle box. From this and other work he was doing, Thorndike postulated the "law of effect," which states that if a response in the presence of a stimulus is followed by a positive event, the association between the stimulus and the response will be strengthened. Conversely, if the response is followed by an aversive event, the association will be weakened. E. L. Thorndike, "Animal Intelligence: An Experimental Study of the Association Processes in Animals," *Psychological Review: Monograph Supplements* 2, no. 4 (1898): entire issue; E. L. Thorndike, *Animal Intelligence: Experimental Studies* (New York: Macmillan, 1911); "marked invasion": J. B. Calhoun and W. L. Webb, "Induced Emigrations among Small Mammals," *Science* 117 (1953): 358–60. In this work, the small mammals were from the genera *Peromyscus*, *Blarina*, *Clethrionomys*, and *Sorex*; "to be an important key": The Theoretical Framework from Which I Approach Problems on Social Behavior, ca. 1953, box 23, folder 14, John B. Calhoun Papers, NLM; "the biology of groups": Job Description for Position at Walter Reed Army Medical Center, 1951, box 23, folder 11, John B. Calhoun Papers, NLM.

3. "No sooner had I arrived": J. Calhoun, "Looking Backward from 'The Beautiful Ones,'" in *Discovery Processes in Modern Biology: People and Processes in Biological Discovery*, ed. W. Klemm (New York: Krieger, 1977), 42; "indicate that it is not an accepted science" and "From such a vantage point": J. Q. Stewart, "Concerning 'Social Physics,'" *Scientific American* 178, no. 5 (May 1948): 20–23, 20; G. K. Zipf, *Human Behavior and the Principle of Least Effort* (Boston: Addison-Wesley, 1949); "The basic particle": J. Calhoun, "The Social Use of Space," in *Physiological Mammalogy*, eds. W. Mayer and R. Van Gelder (New York: Academic Press, 1963), 1–185, 2–3; "of every individual being affected": Calhoun, "Looking Backward," 42; "Since it is logical to assume": J. B. Calhoun, "Behavior of House Mice with Reference to Fixed Points of Orientation," *Ecology* 37, no. 2 (1956): 287–301, 289; "a modicum of mild aggression": J. Calhoun, "What Sort of Box?," *Man-Environment Systems* 3 (1973): 12. Also see J. B. Calhoun, "A Comparative Study of the Social Behavior of Two Inbred Strains of House Mice," *Ecological Monographs* 26 (1956): 81–103; J. Calhoun, "The Study of Wild Animals under Controlled Conditions," *Annals of the New York Academy of Sciences* 51, no. 6 (1950); and J. B. Calhoun, "The Social Aspects of Population Dynamics," *Journal of Mammalogy* 33, no. 2 (1952): 139–59.
4. "As our human society": Calhoun, "Social Aspects," 151; "To spread a layer of human protoplasm": W. Bateson, *Biological Fact and the Structure of Society: The Herbert Spencer Lecture Delivered at the Examination Schools on Wednesday, February 28, 1912* (Oxford: Clarendon Press, 1912); "a common denominator," "The greater the diversity," and "the greater will be the effect": J. Calhoun, "Ecological Principles and the Concept of Morality," 1950, box 6, folder 27, John B. Calhoun Papers, NLM; "didn't think in a linear": Author interview with Catherine Calhoun, July 5, 2022; "sufficed to," "On arriving at," and "he could never get": Calhoun, "Looking Backward," 46, and see also The Theoretical Framework from Which I Approach Problems on Social Behavior, ca. 1953, NLM.

CHAPTER 6

1. "psychiatric casualties": L. Squire, *The History of Neuroscience in Autobiography*, vol. 1 (New York: Academic Press, 1996), 199; "Skinnerian grandchildren": J. Calhoun, "Looking Backward from 'The Beautiful Ones,'" in *Discovery Processes in Modern Biology: People and Processes in Biological Discovery*, ed. W. Klemm (New York: Krieger, 1977), 46; "the interrelated observations": E. Hall, *The Hidden Dimension* (New York: Doubleday, 1966), 1; "encouraged [my] intellectual collaboration": Calhoun, "Looking Backward," 47; "The Korean War was then in full swing": Calhoun, "Looking Backward," 47; "the marked repetition": Calhoun to Rioch, January 15, 1955, box 23, folder 12, John B. Calhoun Papers, NLM; Considerations concerning the Continuance of Studies in Animal Behavior from an Ecological Viewpoint under Army Auspices, 1953, box 23, folder 15, John B. Calhoun Papers, NLM.

2. "began to worry what McCarthy": Calhoun, "Looking Backward," 48; "The area of research," "curtailment of funds," "a nice way of phrasing," "My interest in group dynamics," and "in a sense": Calhoun to Shakow, December 18, 1953, box 23, folder 14, John B. Calhoun Papers, NLM; "at the end of all [his] hopes": Calhoun, "Looking Backward," 47; "Exactly what transpired": Calhoun, "Looking Backward," 48; "senior investigator in the comparative": 1955 Job Description at the NIMH, 1955, box 23, folder 22, John B. Calhoun Papers, NLM. Calhoun's letter of resignation from Walter Reed can also be found in box 23, folder 22.
3. Casey granted the agency a six-year lease: It appears this was a renewal of a lease first signed in 1949, but no action by Casey or NIMH was taken until the renewal began. See box 23, folder 1, John B. Calhoun Papers, NLM, for initial agreement; "what, if any, contributions": Casey to Albert Siepert, December 6, 1955, box 23, folder 21, John B. Calhoun Papers, NLM; Director of NIH and the director of NIMH: Seymour Kety to Robert H. Felix (NIMH) and William Sebrell Jr. (NIH), June 2, 1955, box 23, folder 21, John B. Calhoun Papers, NLM.
4. "complex social group(s)": Proposed project at Rockville farm, 1956, Calhoun, box 23, folder 24, John B. Calhoun Papers, NLM; "living in a small town": Calhoun to Casey, December 2, 1957, box 23, folder 21, John B. Calhoun Papers, NLM; "for a period of": J. Calhoun, Rockville Farm Barn, 1955, p. 7, box 23, folder 21, John B. Calhoun Papers, NLM.
5. "The concept of carrying capacity" and "buffering the individual": J. B. Calhoun, "Social Welfare as a Variable in Population Dynamics," *Cold Spring Harbor Symposia on Quantitative Biology* 11 (1957): 339–55, 339; "on the utilization of time": Calhoun, "Social Welfare as a Variable," 354; J. Watson and F. Crick, "The Structure of DNA," *Cold Spring Harbor Symposia on Quantitative Biology* 18 (1953): 123–31; For more on the space cadets, see E. Ramsden, "Stress in the City: Mental Health, Urban Planning, and the Social Sciences in the Postwar United States," in *Stress, Shock, and Adaptation in the Twentieth Century*, eds. D. Cantor and E. Ramsden (Rochester, NY: University of Rochester Press, 2014), 291–319; E. Ramsden and J. C. Adams, "Escaping the Laboratory: The Rodent Experiments of John B. Calhoun and Their Cultural Influence," *Journal of Social History* 42, no. 3 (2009); Space Cadets, Comments for the Third Conference on Social and Physical Environment Variables as Determinants of Mental Health, 1957, box 56, folder 13, John B. Calhoun Papers, NLM; and L. Duhl, ed., *The Urban Condition* (New York: Basic Books, 1963). For more on Leonard Duhl, see Manuscripts and Other Writings of Leonard J. Duhl, Menninger Foundation Archives, Kansas State University; N. Rashevsky, *Mathematical Biophysics: Physicomathematical Foundations of Biology* (Chicago: University of Chicago Press, 1938); N. Rashevsky, *Advances and Applications of Mathematical Biology* (Chicago: University of Chicago Press, 1940). For more on Rashevsky, see T. Abraham, "Nicolas Rashevsky's Mathematical Biophysics," *Journal of the History of Biology* 37 (2004): 333–85.

CHAPTER 7

1. "for their discoveries," "behaviour patterns become explicable," and "the most eminent founders": Karolinksa Institutet, "Nobel Prize in Physiology or Medicine 1973," press release, https://www.nobelprize.org/prizes/medicine/1973/press-release/; N. Tinbergen, *The Study of Instinct* (Oxford: Clarendon Press, 1951); For more on von Frisch and the waggle dance, see T. Munz, *The Dancing Bees: Karl von Frisch and the Discovery of the Honeybee Language* (Chicago: University of Chicago Press, 2016); S. Kawamura, "The Process of Sub-Culture Propagation among Japanese Macaques," *Primates* 2 (1959): 43–60; "Birds described as tits": J. Fisher and R. Hinde, "The Opening of Milk Bottles by Birds," *British Birds* 42 (1949): 347–57.
2. "corresponds to a possible": G. E. Hutchinson, "Concluding Remarks," *Cold Spring Harbor Symposium* 22 (1957): 415–27, 416. Also see R. Holt, "Bringing the Hutchinsonian Niche into the 21st Century: Ecological and Evolutionary Perspectives," *Proceedings of the National Academy of Sciences of the United States of America* 106 (2009): 19659–65; D. Lack, "The Significance of Clutch-Size," *Ibis* 89, no. 2 (1947): 302–52; D. Lack, "The Significance of Clutch-Size in the Partridge (*Perdix perdix*)," *Journal of Animal Ecology* 16, no. 1 (1947): 19–25; D. Lack, "Clutches above the Normal Limit," in *The Natural Regulation of Animal Numbers* (New York: Oxford University Press, 1954), 31; For more on Lack, see T. Anderson, *The Life of David Lack: Father of Evolutionary Ecology* (New York: Oxford University Press, 2013); W. Thorpe, "David Lambert Lack, 1910–1973," *Biographical Memoirs of the Fellows of the Royal Society of London* 20 (1974): 271–93; and L. S. Forbes, "The Good, the Bad and the Ugly: Lack's Brood Reduction Hypothesis and Experimental Design," *Journal of Avian Biology* 25, no. 4 (1994): 338–43; V. C. Wynne-Edwards, "The Control of Population-Density through Social Behaviour: A Hypothesis," *Ibis* 101, no. 3/4 (1959): 436–41. A few years later, Wynne-Edwards presented his idea in book form as V. C. Wynne-Edwards, *Animal Dispersion in Relation to Social Behaviour* (Edinburgh: Oliver & Boyd, 1962). For more on Wynne-Edwards, see I. Newton, "Vero Copner Wynne-Edwards," *Biographical Memoirs of the Fellows of the Royal Society of London* 44 (1998): 473–84.
3. "cooperate with marked precision": 1960–61 Research Schedule, 1960, box 23, folder 47, John B. Calhoun Papers, NLM; "An innovation by a single rat": J. Calhoun, "What Sort of Box?," *Man-Environment Systems* 3 (1973): 14; "To the [co-operator] rat" and "To correct this behavior": Calhoun, "What Sort of Box?," 14.
4. In three other enclosures, water dispensers were constructed so that it took significant time to access the water and so that many rats could drink at once, while food dispensers contained ground-up pellets and were laid out so that it was likely that rats would engage in relatively little interaction with others.
5. "To do this on a rigorous scientific basis": Calhoun to Casey, December 2, 1957, box 23, folder 21, John B. Calhoun Papers, NLM; "income factor": J. Calhoun,

"A Behavioral Sink," in *Roots of Behavior: Genetics, Instinct, and Socialization in Animal Behavior*, ed. E. Bliss (New York: Harper, 1962), 295–315, 296. For more on the STAW experiment, see Beginning Statement of the 1956–62 Research at the Casey Barn, ca. 1958, box 23, folder 32, John B. Calhoun Papers, NLM. For more on the talk Calhoun planned to give in Denver, see 1960–61 Research Schedule, 1960, NLM.

CHAPTER 8

1. "In the celebrated thesis" and "no escape from the behavioral": J. Calhoun, "Population Density and Social Pathology," *Scientific American* 206, no. 2 (1962): 139; "Phlegmatic animals": Calhoun, "Population Density and Social Pathology," 143. There was one case in which the model failed to predict the distribution of the rats. Calhoun interpreted that case as one in which rat tradition could, at least temporarily, keep them from distributing themselves into neighborhoods as his model predicted. Very early on in a spin-off experiment, just by chance, some of the young rats in the first generation showed a strong preference for neighborhood III (two ramps, but a high-rise building). Quickly, all the other rats spent their time there as well. That preference for neighborhood III continued for many months, even though a low-rise building in neighborhood II (which also had two ramps) was available. The rats clung to their neighborhood III tradition for many months, until eventually redistributing themselves in a manner similar to what Calhoun's model predicted.
2. Calhoun published his results on these enclosure experiments in two papers and one book chapter: J. B. Calhoun, "Determinants of Social Organization Exemplified in a Single Population of Domesticated Rats," *Transactions of the New York Academy of Sciences* 5 (1961): 437–42; Calhoun, "Population Density and Social Pathology"; and J. Calhoun, "A Behavioral Sink," in *Roots of Behavior: Genetics, Instinct, and Socialization in Animal Behavior*, ed. E. Bliss (New York: Harper, 1962). For more on velocity, see J. B. Calhoun, "Social Welfare as a Variable in Population Dynamics," *Cold Spring Harbor Symposia on Quantitative Biology* 11 (1957); J. B. Calhoun, *The Study of Velocity* (n.p.: US Department of Health, Education, and Welfare, 1962) from the collection of Catherine Calhoun; and J. B. Calhoun, "Crowding and Social Velocity," in *New Dimensions in Psychiatry: A World View*, vol. 2, eds. S. Arieti and G. Chrzanowski (New York: John Wiley and Sons, 1977), 20–44.
3. A related experiment, discussed in Calhoun's "A Behavioral Sink," allowed for six generations before the experiment was terminated. Those populations did not crash to extinction; "is not surprising in view": Calhoun, "A Behavioral Sink," 300; "to assure a conditioned social contact" and "It reflects a redefinition": Calhoun, "Determinants of Social Organization," 438–49; "pathological

togetherness": Calhoun, "Population Density and Social Pathology," 139; "several young rats feed simultaneously": Calhoun, "A Behavioral Sink," 300; "His laboratory rats huddle": "News from the Animal Kingdom," *Life*, January 25, 1960; "an estrous female would be" and "These females simply piled": Calhoun, "Population Density and Social Pathology," 145; "if any situation arose": Calhoun, "Population Density and Social Pathology," 144; "exhibited occasional signs of pathology" and "Below the dominant males": Calhoun, "Population Density and Social Pathology," 146; "represents a form of creativity": Calhoun, "Determinants of Social Organization," 440; "were completely passive" and "their social disorientation": Calhoun, "Population Density and Social Pathology,"146; "In spite of the fact": Calhoun, "Population Density and Social Pathology," 141.

4. "The world's population has been growing so fast," "I have two ways of looking," and "a sort of withdrawal": "Overcrowding Caused a Social Breakdown," *Washington Daily News*, September 6, 1964; "The world's population is increasing": "Gruesome Effects Laid to Overcrowding," *Washington Post*, February 10, 1964; "read with keen interest" and "natural check(s) built": Anders Richter to John Calhoun, March 6, 1962, box 138, folder 3, John B. Calhoun Papers, NLM; Professors and graduate students: Author interview with Neil Greenberg, August 1, 2022; Calhoun's *Scientific American* article was also reprinted a few years later in ecologist Garrett Hardin's 1969 collection, *Population, Evolution, and Birth Control: A Collage of Controversial Ideas* (San Francisco: W. H. Freeman); "new unpopulated areas": P. L. Broadhurst, "Anarchy in Rat Town," *Aspect* 8 (1963): 51–58; "Certainly, many species," "as sources of water," and "natural selection will favor": Calhoun, "A Behavioral Sink," 314.
5. "to provide a climate": NIH Behavioral Research Field Station, 1959, box 23, folder 36, John B. Calhoun Papers, NLM; "catwalks," Proposal for a New Facility for Calhoun at the New NIH Farm, 1959, box 23, folder 40, John B. Calhoun Papers, NLM; For more on the NIH Behavior Research Field Station, see NIH Behavioral Research Field Station, 1959, NLM; and NIMH Brain Behavior Lab, 1959, box 23, folder 35, John B. Calhoun Papers, NLM; "to get away": J. Calhoun, "Looking Backward from 'The Beautiful Ones,'" in *Discovery Processes in Modern Biology: People and Processes in Biological Discovery*, ed. W. Klemm (New York: Krieger, 1977), 61; "It is obvious that the": Calhoun, "Population Density and Social Pathology," 148; "I expect to pursue": J. Calhoun, "Concerning the Development of a Point of View: An Autobiography by John B. Calhoun," 1962, box 23, folder 57, John B. Calhoun Papers, NLM.
6. "to increase knowledge": https://casbs.stanford.edu/about/history; "Ecology represents a point of view" and "I will leave to you": Space Cadets, 39th Annual Meeting of American Orthopsychiatric Association: The Environment of the Metropolis, 1962, box 23, folder 35, John B. Calhoun Papers, NLM.

CHAPTER 9

1. I. DeVore, ed., *Primate Behavior: Field Studies of Monkeys and Apes* (New York: Holt, Rinehart, and Winston, 1965); "Just solitude if one wished it," "odd-ball, on-the-ball," and "three hours and many equations later": Comments on 1962–63 Fellowship at Center For Advanced Study in the Behavioral Sciences, Stanford, California, 1964, box 23, folder 76, John B. Calhoun Papers, NLM. For more on Mosteller, see F. Mosteller, "The World Series Competition," *Journal of the American Statistical Association* 47, no. 259 (1952): 355–80.
2. "page after page": J. Calhoun, "Looking Backward from 'The Beautiful Ones,'" in *Discovery Processes in Modern Biology: People and Processes in Biological Discovery*, ed. W. Klemm (New York: Krieger, 1977), 60; C. R. Rogers, *On Becoming a Person: A Therapist's View of Psychotherapy* (Boston: Houghton Mifflin, 1961); "The resemblance to human behavior": C. R. Rogers, "Some Social Issues Which Concern Me," *Journal of Humanistic Psychology* 12, no. 2 (1972): 45–60, 49; "informal discussions": J. B. Calhoun, *The Study of Velocity* (n.p.: US Department of Health, Education, and Welfare, 1962) from the collection of Catherine Calhoun. For Calhoun's work on mass panic, see J. Calhoun, "Induced Mass Movements of Small Mammals: A Suggested Program of Study," 1962, collection of Catherine Calhoun, written while in residence at CASBS 1962; "switched [him] more strongly": Calhoun, "Looking Backward," 62; "represent only crude approximations to reality": J. Calhoun, "The Social Use of Space," in *Physiological Mammalogy*, eds. W. Mayer and R. Van Gelder (New York: Academic Press, 1963), 184–85; "above or below which": Calhoun, "Social Use of Space," 2; "Development of a larger": Calhoun, "Social Use of Space," 101; "Human society has developed": Calhoun, "Social Use of Space," 184; "reorientation of the value system": Calhoun, "Social Use of Space," 101; "without knowledge of evolutionary": Calhoun, "Social Use of Space," 184; "Portions of Calhoun's interesting": B. Welch, "Experimental Population Ecology," *Science* 146 (1964): 49; "empirical observations with rats": D. D. Thiessen, "Review of *Physiological Mammalogy*," *Human Biology* 37, no. 2 (1965): 206–8, 206; "As the result of some": W. R. Eadie, "Review of *Physiological Mammalogy*," *Quarterly Review of Biology* 40 (1965): 106–7, 106.
3. "the time of Christ": H. von Foerster, P. Mora, and L. Amiot, "Doomsday: Friday, 13 November, A.D. 2026," *Science* 132 (1960): 1291–95, 1293; "to infinity": von Foerster et al., "Doomsday," 1292; "Dr. von Foerster is the father of three": "Science: Doomsday in 2026 A.D.," *Time*, November 14, 1960; "The logic and beauty" and "The increasingly fragmented": Calhoun, "Looking Backward," 62. Calhoun's underlined version of the doomsday paper can be found in Misc. Docs re: 317 P.H., 1962–1968, box 45, folder 41, John B. Calhoun Papers, NLM; "I could not help but ask" and "capacity as scientist": Comments on 1962–63 Fellowship at Center for Advanced Study, 1964, NLM.

4. J. B. Calhoun, "A: A Mathematical Construct, 1962 Poem," 1962, collection of Catherine Calhoun, written at CASBS; "direct selection toward," "flattened split nostrils," "achieved the capacity," and "internal cybernetic detoxification system": 317 P.H. (Posthomo) a Satire on a Future Multiple "Utopia," 1963–1964, box 45, folder 38, John B. Calhoun Papers, NLM. Also see Misc. Docs re: 317 P.H., 1962–1968, NLM; "Although I have filled notebooks": Calhoun, "Looking Backward," 62. The now long-debunked claim that humans had gone through an aquatic phase in recent evolutionary history had been floating around in the anthropological literature since Alister Hardy's 1960 paper "Was Man More Aquatic in the Past?" (*New Scientist* 7: 642–45), and would be popularized in Desmond Morris's 1967 book, *The Naked Ape: A Zoologist's Study of the Human Animal* (New York: McGraw-Hill).
5. "Studies are widely referred": John Calhoun Curriculum Vitae, 1963–1983, box 1, folder 1, John B. Calhoun Papers, NLM. For more on Calhoun's trip to Africa, see Travel, 1963, box 23, folder 67, John B. Calhoun Papers, NLM. For more on McHarg and Calhoun, see University of Pennsylvania, 1963–1968, box 6, folder 49, John B. Calhoun Papers, NLM, which includes McHarg's invitation for a return visit; G. Hardin, *Population, Evolution, and Birth Control: A Collage of Controversial Ideas* (San Francisco: W. H. Freeman, 1969). For a schematic of the chicken experiment, see Velocity Study Pens (Chickens), 1963, box 45, folder 15, John B. Calhoun Papers, NLM.

CHAPTER 10

1. "it has been said": Calhoun Lab Progress Report, 1969–1970, box 17, folders 15–16, John B. Calhoun Papers, NLM.
2. R. Dubos, *So Human an Animal: How We Are Shaped by Surroundings and Events* (New York: Charles Scribner and Sons, 1968); "a modified version of enclosures" and "unusual aggressiveness": A. Kessler, *Interplay between Social Ecology and Physiology, Genetics and Population Dynamics of Mice* (PhD diss., Rockefeller University, 1966), 8; "standing room only": J. Calhoun, "What Sort of Box?," *Man-Environment Systems* 3 (1973): 23; "In a world beset": Calhoun Lab Progress Report, 1969–1970, NLM. In early 1966, Calhoun convened an NIMH-sponsored workshop on "experimental universes," both to discuss this general approach to studying population growth and to get as much feedback as possible on his ideas for his mice and rats. One of the participants at the workshop was Alexander Kessler.
3. "Everything . . . was to be Utopian" and "euphoric in planning": Calhoun, "What Sort of Box?," 24. In some universes there were two water bottles per cell.
4. "sound mental health planning," "standardization, order, predictability," "programmed diversity, programmed uncertainty," and "I am approaching 50": 1965

Space Cadets, Conference, 1965, box 59, folders 3 and 4, John B. Calhoun Papers, NLM.

5. J. Calhoun, "The Role of Space in Animal Sociology," *Journal of Social Issues* 22, no. 4 (1966): 46–59; "understanding, creativity": J. Calhoun, "A Glance into the Garden," in *Three Papers on Human Ecology*, ed. D. Bowers (Oakland, CA: Mills College Assembly Series, 1966), 19–36, 33; "our fellow man," "a network of communication," and "My studies on rodents": Calhoun, "Glance into the Garden," 32; "My studies on rodents": Calhoun, "Glance into the Garden," 29; "unofficial directory for a new decade," "a directory of the invisible colleges," and "Sociology, Psychology": Unofficial Directory for a New Decade, 1970, box 8, folder 59, John B. Calhoun Papers, NLM; "constructive sociology": H. G. Wells, *World Brain* (New York: Doubleday, Duran and Co., 1938), 5; Clarke lists the year 2100 in a table on page 223 of A. Clarke, *Profiles of the Future* (London: Victor Gollancz, 1962). For more on Vickers, see G. Vickers, "The Psychology of Policy Making and Social Change," *British Journal of Psychology* 110, no. 467 (1964): 465–77. It was Yale historian of science Derek J. de Solla Price who noted that the term "invisible college" could be traced back to the 1600s. Price also discussed invisible colleges in the 1960s, but they were at a smaller scale than what Calhoun envisioned and used journals like *Nature* to connect to one another. D. Price, "The Scientific Foundations of Science Policy," *Nature* 206 (1965): 233–38.
6. "normal social pressures," "little awareness of," "This formulation of creativity," "Although my observations," and "Upon first inspection": Calhoun, "Behavioral States and Developed Images," (lecture: American Association for the Advancement of Science, Dallas, TX, 1965), pages 7, 6, 7, 6, and 15, respectively; "from this firmer path": Calhoun, "Behavioral States and Developed Images," 1–2; "Men and institutions": Calhoun, "Population and Mental Health Progress," 1966, item 3, collection of Catherine Calhoun.
7. "the sociological and psychological aspects" and "a deterioration of the relations": Calhoun, "Animal Behavior and Ecology Implications for Planning the Human Scene," 1966, collection of Catherine Calhoun; "Calhoun in his rat": M. Calderone, "Human Cost Accounting," in *The Complete Book of Birth Control* (Los Angeles: American Art Agency, 1965), 5–9, 9.
8. "philosopher, literary critic, historian": "Lewis Mumford, a Visionary Social Critic, Dies at 94," *New York Times*, January 28, 1990; "no small part": L. Mumford, *The Urban Prospect* (New York: Basic Books, 1968), 210; "slurban explosion": Mumford, *The Urban Prospect*, 4; "one of the few extraordinary books": Review of *The Hidden Dimension*, by Edward Hall, *Chicago Tribune*, June 26, 1966; "sufficiently startling to warrant": E. Hall, *The Hidden Dimension: An Anthropologist Examines Man's Use of Space in Public and in Private* (New York: Doubleday, 1966), 24; "The findings of these studies": Hall, *Hidden Dimension*, 25.
9. J. Calhoun, text of speech presented at the 1968 American Association for the Advancement of Science annual meeting, 1968, collection of Catherine Calhoun.

In 1971, Calhoun modified the text, added new results, and published it: J. B. Calhoun, "Space and the Strategy of Life," in *Behavior and Environment: The Use of Space by Animals and Men*, ed. A. H. Esser (New York: Plenum Press, 1971), 329–87. All quotes here are from the 1968 speech: "general tension accompanying" (p. 2), "Whether this family history" (p. 2), "standing room only" (p. 6), "All of these objects" (p. 6), "Two individuals could thus" (p. 6), "for all practice typical purposes" (p. 6), "the overflow of stressful situations" (p. 7), "To the extent that these insights" (p. 11), "required an augmented awareness" (p. 31), "conceptual homunculus" (p. 29), and "The body of man" (p. 29).

10. "We become linked," "To continue enlarging conceptual," "more than 99%," and "one hundred thinking protheses,": American Association for the Advancement of Science speech, 1968, pages 35, 24, 35, and 35, respectively; Szilard wrote "Calling All Stars" in 1949. The original, typed manuscript can be found here: https://library.ucsd.edu/dc/object/bb67934053/_1.pdf. *Calling All Stars* was part of a 1961 collection of Szilard's stories, *The Voice of the Dolphins and Other Stories* (New York: Simon and Schuster). Page numbers here refer to the 1961 collection: "observed on the earth" (p. 107) and "Our society consists of 100 minds" (p. 105); "important topic": Chastain to Calhoun, October 22, 1968, box 8, folder 44, John B. Calhoun Papers, NLM; "I make progress": Calhoun to Chastain, February 5, 1969, box 8, folder 44, John B. Calhoun Papers, NLM; "a new type of": J. B. Calhoun, "The Positive Animal: Increased Human Potentiality Enhances Stability of the Total Ecosystem and Preserves Evolution," *Man-Environment Systems* (September 1971): 1–5.

CHAPTER 11

1. C. Darwin, *The Various Contrivances by which Orchids Are Fertilised by Insects* (London: Charles Murray, 1862); Darwin received the *Angraecum sesquipedale* plant from James Bateman. "I have just received": Darwin to J. D. Hooker, January 25, 1862, Darwin Correspondence Project, University of Cambridge, https://www.darwinproject.ac.uk/letter/DCP-LETT-3411.xml. For more on the discovery of *Xanthopan morganii praedicta*, see J. Arditti, J. Elliott, I. J. Kitching, and L. T. Wasserthal, "'Good Heavens What Insect Can Suck It'—Charles Darwin, *Angraecum sesquipedale* and *Xanthopan morganii praedicta*," *Botanical Journal of the Linnean Society* 169, no. 3 (2012): 403–32; and C. Netz and S. S. Renner, "Long-Spurred *Angraecum* Orchids and Long-Tongued Sphingid Moths on Madagascar: A Time Frame for Darwin's Predicted *Xanthopan/Angraecum* Coevolution," *Biological Journal of the Linnean Society* 122, no. 2 (2017): 469–78; "a new adaptive zone": P. Ehrlich and P. Raven, "Butterflies and Plants: A Study in Coevolution," *Evolution* 18, no. 4 (1964): 586–608, 591.
2. Ehrlich credits the title of his book to the cover of a pamphlet produced by the Hugh Moore Fund for International Peace; "*Population Control or Race to*

Oblivion?": P. Ehrlich, *The Population Bomb* (New York: Ballantine Books and the Sierra Club, 1968), opening page; "virtually a co-author": Ehrlich, *Population Bomb*, 223; "Overpopulation is now": Ehrlich, *Population Bomb*, opening page; "The birth rate," "population control at home," and "responsibility prizes": Ehrlich, *Population Bomb*, prologue; "It absolutely fascinated me": Author interview with Steve Suomi, July 26, 2022; "We know all too well that when rats" and "extrapolating from the behavior of rats": Ehrlich, *Population Bomb*, 168.; "just possibilities, not predictions": Ehrlich, *Population Bomb*, 7; "in the 1970's the world will undergo famines": Ehrlich, *Population Bomb*, prologue; "The biggest tactical error": P. Ehrlich and A. Ehrlich, "*The Population Bomb* Revisited," *Electronic Journal of Sustainable Development* 1, no. 3 (2009): 63–71, 66. Also see P. Ehrlich and J. P. Holdren, "Impact of Population Growth: Complacency concerning This Component of Man's Predicament Is Unjustified and Counterproductive," *Science* 171 (1971): 1212–17.

3. W. Sullivan, "An Attack on Man the Aggressor; Scientists in Rush to Study the Pattern of Destruction," *New York Times*, August 26, 1968; T. Wolfe, *The Pump House Gang* (New York: Farrar, Straus & Giroux, 1968); T. Wolfe, *The Electric Kool-Aid Acid Test* (New York: Farrar, Straus & Giroux, 1968). "Oh Rotten Gotham! Sliding Down into the Behavioral Sink" was first published in 1966 in the *New York World Journal Tribune*; "I just spent two days" and "It got to be easy": Wolfe, *Pump House Gang*, 295; "In one major experiment": Wolfe, *Pump House Gang*, 298; "Two dominant male rats": Wolfe, *Pump House Gang*, 298–300; "Ned [Edward Hall] and I share the view": J. Calhoun, "Looking Backward from 'The Beautiful Ones,'" in *Discovery Processes in Modern Biology: People and Processes in Biological Discovery*, ed. W. Klemm (New York: Krieger, 1977), 47; "The term itself": Hunter Thompson to Tom Wolfe, April 21, 1968, in D. Brinkley, ed., *Fear and Loathing in America: The Brutal Odyssey of an Outlaw Journalist, The Gonzo Letters, Part II, 1968–1976* (New York: Simon & Schuster, 2000), 54; Also see I. Kristol, "It's Not a Bad Crisis to Live In," *New York Times*, January 22, 1967.
4. I. McHarg, *Design with Nature* (New York: American Museum of Natural History, 1969). For more on McHarg, see S. Herrington, "The Nature of Ian McHarg's Science," *Landscape Journal* 29, no. 1 (2010): 1–20; J. C. Adams and E. Ramsden, "Rat Cities and Beehive Worlds: Density and Design in the Modern City," *Comparative Studies in Society and History* 53, no. 4 (2011): 722–56; and B. Fleming, "50 Years after *Design with Nature*, Ian McHarg's Ideas Still Define Landscape Architecture," *Metropolis*, June 18, 2019, https://metropolismag.com/viewpoints/mcharg-design-with-nature-50th-anniversary/; *Multiply and Subdue the Earth* (National Educational Television and Radio Center Producing Organization and WGBH Educational Foundation, 1969).
5. "These poor beautiful mice never learn," "blobs of protoplasm," "frozen in a child-like trance," "line up, on the low elevation," and "They are deathly quiet": T. Huth, "Of Mice and Men," *Washington Post*, March 29, 1971; "with

little capacity": Progress Report on Environmental and Genetic Variables Affecting Biological Systems, July 1, 1970 through June 30, 1971, box 17, folders 15–16, John B. Calhoun Papers, NLM. For more on the mice universes at this time, see NIMH Progress Reports on Environmental and Genetic Variables Affecting Biological Systems, July 1, 1968 through June 30, 1969, and July 1, 1967 through June 30, 1968, box 17, folders 15–16, John B. Calhoun Papers, NLM; "would collect about any strange object," *Looking Forward*, September 17, 1992, box 138, folder 25, John B. Calhoun Papers, NLM.

6. J. B. Calhoun, "Design for Mammalian Living," *Architectural Association Quarterly* 1 (1969): 24–35; "concerned with bricks and mortar," "enhancing creativity," and "increasing valuing": J. B. Calhoun, "Meta-Environmentalism," *Man-Environment Systems* (July 1969): 103, reprinted in Congressional Record, Proceeding and Debates of the 92nd Congress, First Session, vol. 117, no. 47, Thursday, April 1, 1971, S4404; "when channels of contact": J. Calhoun, Thoughts following the Conservation Foundations "conversations" of May 20 and 21, 1969, on population and the environment, 1969, collection of Catherine Calhoun.
7. R. Ardrey, *The Territorial Imperative: A Personal Inquiry into the Animal Origins of Property and Nations* (New York: Atheneum, 1966); "A society is a group of unequal beings," "equality of opportunity," and "The just society": R. Ardrey, *The Social Contract: A Personal Inquiry into the Evolutionary Sources of Order and Disorder* (New York: Atheneum Publishers, 1970), 3; "horrifying human implications": Ardrey, *Social Contract*, 217; "a maverick's maverick": Ardrey, *Social Contract*, 183–84; "more ominous experiments": Ardrey, *Social Contract*, 216; "The three sciences central": Ardrey, *Social Contract*, 21; "lack of coherence," "an excellent raconteur," and "misanthropic": C. L. Brace, review of *The Social Contract*, by Robert Ardrey, *American Scientist* 59, no. 3 (1971): 376–77; "jewels . . . mixed with drek": M. H. Fried, review of *The Social Contract*, by Robert Ardrey, *American Journal of Sociology* 77, no. 1 (1971): 149–53; "inaccuracy or intellectual sloppiness": H. Callan, review of *The Social Contract*, by Robert Ardry, *American Anthropologist* 74, no. 6 (1972): 1536–38; "Ardrey is a child": Comments upon Reading the Uncorrected Proofs of Robert Ardrey's "The Social Contract," 1970, box 9, folder 9, John B. Calhoun Papers, NLM, 12–14; R. Linton, *Terracide: America's Destruction of Her Living Environment* (Boston: Little, Brown, 1970); "for Adult Intellectuals only": S. Rodriguez, K. Deitch, S. C. Wilson, J. Green, R. Hayes, J. Osborne, R. Brand, and W. Mendes, *Insect Fear: Tales from the Behavior Sink* no. 2, 1970. Ramsden and Adams, in "Escaping the Laboratory," call *Insect Fear* "a garish, Robert Crumb-meets-William Burroughs" and suggest that the writers likely learned of Calhoun's behavioral sink through Tom Wolfe's *New York World Journal Tribune* article, "Oh Rotten Gotham!"
8. "Neither: rather we approach" and "The tactic is to describe": Calhoun, A Strategy of Inquiry, collection of Catherine Calhoun; "except for a few" and "never had the opportunity to": Progress Report on Environmental and Genetic

Variables, July 1, 1970 through June 30, 1971, NLM; "between twenty-two hundred and twenty-five hundred mice": J. Calhoun, "John B. Calhoun Film 7.1 [edited], (NIMH, 1970-1972)," National Library of Medicine, uploaded November 9, 2017, YouTube video, 38:28, https://www.youtube.com/watch?v=iOFveSUmh9U, at 19:40.

9. "It's a lovely day," "a smallish, cheerful man," "All this was familiar," "a strong impulse," "the rodential Bourgeoisie," "The difference was obvious," "Aren't we maybe seeing," and "This generation of the young": all from S. Alsop, "Dr. Calhoun's Horrible Mousery," *Newsweek*, August 17, 1970, 96; "I was fascinated": Julie Lach to Calhoun, March 30, 1971, box 9, folder 27, John B. Calhoun Papers, NLM; "It causes me considerable": Calhoun to Lach, April 7, 1971, box 9, folder 27, John B. Calhoun Papers, NLM.

CHAPTER 12

1. "Nothing [she] had heard": R. O'Brien, *Mrs. Frisby and the Rats of NIMH* (New York: Atheneum Books, 1971), book jacket; "He remembers" and "In the story": S. Rovner, "Rats! The Real Secret of NIMH," *Washington Post*, July 21, 1982; her father did visit: J. Conly, "Intelligence and Utopia in *Mrs. Frisby and the Rats of NIMH*," in *Philosophy in Children's Literature*, ed. P. Costello (Lanham, MD: Lexington Books, 2012), 204; "The Plan of the Rats of NIMH": O'Brien, *Mrs. Frisby*, 82; "Did O'Brien get the London": Calhoun's notes on *Mrs. Frisby and the Rats of NIMH* and *The Secret of NIMH*, n.d., collection of Catherine Calhoun. For more on Robert O'Brien, see A. R. Silver, "The Book of Nicodemus and Other Apocrypha: The Works of Robert C. O'Brien as a Reflection of Technological/Scientific Anxieties in 1960s American Culture" (MA thesis, University of Michigan–Flint, 2019); the NBC outreach comes from Calhoun to Jun-Ichiro Takeda, October 15, 1980, Media requests, 1980, box 15, folder 42, John B. Calhoun Papers, NLM.
2. "(A) critical aspect of the population problem" and "Dr. Calhoun is careful": Congressional Record, Proceeding and Debates of the 92nd Congress, First Session, vol. 117, no. 47, Thursday, April 1, 1971, S4398. In addition to the *Scientific American* article, the articles entered into the Congressional Record were: J. B. Calhoun, "Meta-Environmentalism," *Man-Environment Systems* (July 1969): 103; and J. B. Calhoun, "The Lemmings' Periodic Journeys Are Not Unique," *Smithsonian Magazine* 1 (1971): 6–13; "Most of them are withdrawn," "a psychologist, philosopher, economist, mathematician," "to conventional science," and "so sometimes we make errors": T. Huth, "Of Mice and Men," *Washington Post*, March 29, 1971; "I had ceased publishing": J. Calhoun, "Looking Backward from 'The Beautiful Ones,'" in *Discovery Processes in Modern Biology: People and Processes in Biological Discovery*, ed. W. Klemm (New York: Krieger, 1977), 53. In "Looking Backward," Calhoun notes that, as far back as the late 1940s, he was not happy with

the whole process: "considerable pressure was put on me to publish the results as a series of short papers" (40).

3. "holistic approach to the interaction": Laboratory of Brain Behavior and Evolution Dedication, Calhoun Speech, 1971, box 25, folder 15, John B. Calhoun Papers, NLM.
4. "unimposing metal prefab": N. Laserson, "It's Not Every Day You Walk into a Laboratory Whose Mission Is to Save the World," *Innovation Magazine* 25 (1971): 14–25, 14; "So people are sitting up": Laserson, "It's Not Every Day," 25; "The inhabitants of [this] universe," "suspect[ed] something similar," and "Buddhism preaches against": Y. Shimuza, "Warning against Overpopulation," *Mainichi Shimbun*, March 1, 1971; "Overpopulation," *Der Spiegel*, May 2, 1971; "Of Mice and Men," *Svenska Dagbladet*, June 12, 1971.
5. "Lower Death Rate in Tested Mice Eventually Stopped Reproduction," *Globe and Mail*, November 2, 1971; "if the progression continues" and "may still face a psychic fate": O. Friedrich, "Population Explosion: Is Man Really Doomed?," *Time*, September 13, 1971.
6. "in an effort to determine," "During this decline," "uncrowded environments," "recoup normal behaviors," and "Our objective here": Progress Report on Environmental and Genetic Variables Affecting Biological Systems, July 1, 1970 through June 30, 1971, box 17, folders 15–16, John B. Calhoun Papers, NLM; "All capacities for developing," "recovered only minimal," and "The only animals that have been": M. Pine, "How the Social Organization of Animal Communities Can Lead to a Population Crisis Which Destroys Them," in *NIMH's Mental Health Program Reports–5*, ed. J. Segal (n.p.: Department of Health, Education, and Welfare, 1971), 158–73, 168.
7. "Our objective here" and "the theoretical origins": Progress Report on Environmental and Genetic Variables Affecting Biological Systems, October 1, 1981 through September 30, 1982, box 17, folders 15–16, John B. Calhoun Papers, NLM; "This technology" and "We believe our use": Calhoun, *Looking Forward*, September 17, 1992, box 138, folder 25, John B. Calhoun Papers, NLM, 21; "carrying such a miniaturization": J. Calhoun, "What Sort of Box?," *Man-Environment Systems* 3 (1973): 22.
8. K. Menninger, *The Crime of Punishment* (New York: Penguin Books, 1968); "overcrowded, unsanitary," "Mental health research," and "he refused to relate his mice": W. Claiborne, "Experts Get Look at – and Smell of – DC Jail," *Washington Post*, October 17, 1971; "you are small and insignificant" and "a situation in which man": J. B. Calhoun, A Collection of Depositions for the DC Jail Case, 1971–1975, collection of Catherine Calhoun; P. Paulus, *Prison Crowding: A Psychological Perspective* (New York: Springer-Verlag, 1988); "if there is a common human condition": R. Wener, *The Environmental Psychology of Prisons and Jails: Creating Humane Spaces in Secure Settings* (New York: Cambridge University Press, 2012); "As

soon as you looked up": Author interview with Freedman, September 13, 2022; J. Freedman, *Crowding and Behavior: The Psychology of High-Density Living* (New York: Viking Press, 1975). Based on studies in animals and humans, Freedman expected that increased density would result in poorer performance on cognitive tests and was "startled" (Freedman, 80) when he found no correlation. Others were as well: his results led to "a collective gasp, similar to the one that must have occurred when a child pointed out that the emperor had no clothes." J. A. Russell and L. Ward, "Environmental Psychology," *Annual Review of Psychology* 33 (1982): 651–88, 673; O. Galle, W. Gove, and J. McPherson, "Population Density and Pathology: What Are the Relations for Man?," *Science* 176 (1972): 23–30.

9. "Each of us," "Dr. Calhoun sees," "Males are painted," and "does not fit the picture": F. Sartwell, "The Small Satanic Worlds of John B. Calhoun," *Smithsonian Magazine* 1 (1970): 69–71, 69; "As a prophet": Sartwell, "Small Satanic Worlds," 71; "Long before the lemmings": J. B. Calhoun, "The Lemmings' Periodic Journeys Are Not Unique," *Smithsonian Magazine* 1 (1971): 6–13, 6; "Author plays Pied Piper," "seems doomed," and "Whenever we fail to produce ideas": Calhoun, "Lemmings' Periodic Journeys," 7; "As each strip of [his] animals," "Perhaps you have heard," "hollow shells," and "In [a] last frenzy of reproduction": Calhoun, "Lemmings' Periodic Journeys," 11. Also see "What Is a Lemming?," *Journal of the American Medical Association* 215 (1971): 1317; "Periodically [man] has been faced" and "Sometimes the new ways": Calhoun, "Lemmings' Periodic Journeys," 12. For more on Calhoun's writings on compassionate revolutions, appreciative systems, and electronic prostheses, see J. B. Calhoun, "Transitory Population Optima in the Evolution of Brain," in *Is There an Optimum Level of Population?*, ed. S. F. Singer (New York: McGraw-Hill, 1971), 265–71; J. B. Calhoun, "Psycho-Ecological Aspects of Population," in *Environ/Mental: Essays on the Planet as a Home*, eds. P. Shepard and D. McKinley (Boston: Houghton Mifflin, 1971); J. B. Calhoun, "The Positive Animal: Increased Human Potentiality Enhances Stability of the Total Ecosystem and Preserves Evolution," *Man-Environment Systems* (September 1971): 1–5; and J. B. Calhoun, "Control of Population: Numbers," *Annals of the New York Academy of Sciences* 184 (1971): 148–55. For more on lemmings and mass suicides in animals, see E. Ramsden and D. Wilson, "The Suicidal Animal: Science and the Nature of Self-Destruction," *Past & Present* 224, no. 1 (2014): 201–42.

CHAPTER 13

1. "The National Institute of Mental Health," "spirit of; essence of," "the spirt and purpose," and "those who share": J. Calhoun, "The Population Crisis Leading to the Compassionate Revolution and Environmental Design," *World Journal of Psychosynthesis* 4 (1972): 21–28, 27; "worldwide communication": Calhoun, "Population Crisis," 26; "aeronautical and space systems": Tentative Agenda: Forum on

Information-Interactions in the Next Generation, May 5–8, 1972, NASA Conference-Brunswick, G.A., 1972, box 11, folder 38, John B. Calhoun Papers, NLM; "idea generator": John Calhoun Curriculum Vitae, 1963–1983, box 1, folder 1, John B. Calhoun Papers, NLM; F. Pohl and C. M. Kornbluth, *The Space Merchants* (New York: Ballantine Books, 1953); A. C. Clarke, "The Sentinel," originally published as "Sentinel of Eternity," in *10 Story Fantasy* (New York: Avon Periodicals, 1951).

2. "Our aim today": J. Z. Young, "Introductory Remarks," *Proceedings of the Royal Society of Medicine* 66 (1973): 79–80; All of the following are from J. Calhoun, "Death Squared: The Explosive Growth and Demise of a Mouse Population," *Proceedings of the Royal Society of Medicine* 66 (January 1973): 80–88: "I shall largely speak of mice," "This takes us back to," "I saw . . . a pale horse" (italics in original), and "Let us first consider the second death" (p. 80); "utopian environment constructed for mice" (p. 81); and "considerable social turmoil" (p. 82); "good maternal care," "number [was] far greater," and "[those] who failed" (p. 84); "transported their young," "started independent life," "for all practical purposes," and "Projection of the prior few months" (p. 85); "At final editing of this paper" (p. 83); "beautiful ones . . . capable only of the most" (p. 86); "about a million observations," "despite the thousandfold increase," and "there was a breaking point" (p. 87); "interrupted Dr Calhoun" and "thought this to be a major trap" (p. 88); "standard argument" and "wild species suffer": F. Sartwell, "The Small Satanic Worlds of John B. Calhoun," *Smithsonian Magazine* 1 (1970): 69; "Balb/C's were aggressive": C. H. Southwick and L. H. Clark, "Aggressive Behaviour and Exploratory Activity in Fourteen Mouse Strains," *American Zoologist* 6 (1966): 559; "May be Year of Doom," *Daily Mirror*, June 23, 1972; C. Dover, "Mice Point the Way to Doom in 1984, Says Scientist," *Daily Telegraph*, June 23, 1973; "Sad Elderly Mice," *Daily Telegraph*, June 23, 1972; "the newspaper death notice" and "You can't identify with nothing": T. Huth, "Ten Dead Boxes of Mice Could be Us. Is Modern Mankind Becoming a Giant Colony of Mice?," *Washington Post*, February 8, 1973; "Mr. Paul Block of *The Tonight Show*": Media requests, 1973, box 13, folder 38, John B. Calhoun Papers, NLM; F. Carey, "Of Mice and Men and Paradise Lost," *Boston Evening Globe*, March 1, 1973; "Easy Living in a High Rise, but Paradise was Brief," *Los Angeles Times*, March 2, 1973; "Our Town: Man or Mouse," *Washington Star-News*, January 11, 1973; "your Universe 25": Carl Bajema to Calhoun, regarding Bajema and Garrett Hardin's textbook, *Biology: Its Principles and Implications*, June 18, 1973, box 4, folder 18, John B. Calhoun Papers, NLM; "Results of Overcrowding," *Good Housekeeping*, June 1973.
3. "Strike No. 1: The shift," "They were suddenly crowded," and "The chairman admonished": J. Calhoun, "Plight of the Ik and Kaiadilt Is Seen as a Chilling Possible End for Man," *Smithsonian Magazine* 3 (1972): 27–32, 29; "Strike No. 1 against these mice" and "all true 'mousity' was lost": Calhoun, "Plight of the Ik," 30; C. Turnbull, *The Mountain People* (New York: Simon and Schuster, 1972). Critiques of Turnbull's work include T. O. Beidelman, review of *The Mountain*

People, by Colin M. Turnbull, *Africa: Journal of the International African Institute* 43, no. 2 (1973): 170–71; F. Barth, "On Responsibility and Humanity: Calling a Colleague to Account," *Current Anthropology* 15 (1974): 99–102; B. Heine, "The Mountain People: Some Notes on the Ik of North-Eastern Uganda," *Africa: Journal of the International African Institute* 55, no. 1 (1985): 3–16; and J. Knight, "'The Mountain People as Tribal Mirror,'" *Anthropology Today* 10, no. 6 (1994): 1–3; "Dick Wakefield in 1972": Collection of Catherine Calhoun. For more on Wakefield, see G. F. Vaughn, "Sheffield's Richard P. Wakefield: Advocate for Human Values, World Futures and the Environment," *Historical Journal of Massachusetts* 32, no. 2 (2004), https://www.westfield.ma.edu/historical-journal/wp-content/uploads/2018/06/Vaughn-summer-2004-combined.pdf.

4. "extremely important": Eleonora Masini to Calhoun, September 5, 1972, Rome 73 correspondence, 1973, box 12, folder 52, John B. Calhoun Papers, NLM; "We are aware of the general," "in this field the Church," and "We wish you well in your work": "Address of the Holy Father Paul VI to the Participants in the Special World Conference on Futures Research, September 27, 1973," in *Insegnamenti di Paolo VI*, vol. XI (Vatican City: Libreria editrice vaticana, 1973), 896–97, Ora n.41 p. 2; "sources with an interest": "Human Needs, New Societies and Supportive Technologies: A Document on the Work of the Groups at the Rome Special World Conference on Futures Research," Project Irades, no. 30 (1973): 35; "The earth is now involved": J. B. Calhoun, "Metascientific Research: A Proposal for a Demonstration Effort to Evaluate 'Research on Synergistic Research' as an Organizational Methodology for Promoting Scientific ℞evolutions in the Understanding of the Complex Dilemmas Characterizing Biomedical-Behavioral-Environmental Phenomena in Contemporary Society," in *Human Needs, New Societies, Supportive Technologies, Collected Documents Presented at the Rome Special World Conference on Futures Research*, vol. III (Rome: Institute of Research and Education in Futures Studies, 1974), 128–57, 129; "from the heights of this villa," "If the 13," "visionary declaration of ℞," "increase the effective contribution," and "its possible role will be explored": A Villa Tuscolana Statement of Intent to Initiate "The Four Catalytic Years" 1974–1977, Rome 73 correspondence, 1973, box 12, folder 52, John B. Calhoun Papers, NLM. For more on correspondence on the Special World Conference on Futures Research, see Calhoun to Eleonora Masini, November 9, 1972, and Masini to Calhoun, November 30, 1972, Rome 73 correspondence, 1973, box 12, folder 52, John B. Calhoun Papers, NLM.

5. "examination of the time course," "aware of less," "beyond this nonchoice," "accompanied by a pervading," and "choose to design further evolution": J. B. Calhoun, "Revolution, Tribalism, and the Cheshire Cat: Three Paths from Now," *Technological Forecasting and Social Change* 4, no. 3 (1973): 263–82, 263; "Our success in being human": Calhoun, "Revolution," 275; "I for one would not like to be a lung fish": J. Calhoun, "What Sort of Box?," *Man-Environment Systems* 3 (1973): 29; "the amount of eccentricity": J. S. Mill, *On Liberty* (London: Longman,

Roberts, & Green Co, 1859), 120–21; "A number of the 'effects'" and "the opportunity to have my consultants": Barry Goldwater Jr. to Calhoun, March 19, 1973, box 4, folder 16, John B. Calhoun Papers, NLM; "although our research," Calhoun to Goldwater Jr., April 6, 1973, box 4, folder 16, John B. Calhoun Papers, NLM.

6. "adapt 'Death Squared' in the development": Calhoun to Karen Hollweg, July 17, 1973, box 14, folder 48, John B. Calhoun Papers, NLM; "Because there are too many people": A. H. Drummond, *The Population Puzzle: Overcrowding and Stress among Animals and Men* (Reading, MA: Addison-Wesley Publishing, 1973), 14; All of the quotes here are from Drummond's *The Population Puzzle*: "man may doing to himself" (pp. 15–16), "Let's go back and look" (p. 40), "a wealth of material" (acknowledgments), "male rats harassed" (p. 44), "a place where undesirable" (p. 50), "happen to overcrowded" (p. 51), "after reading about Calhoun's experiments" (p. 51), "evolved the ability" (p. 123), and "Is [Calhoun] a prophet" (p. 113).

CHAPTER 14

1. "I want to talk," "Accompanying each there was a message," "as environments changed," "We, too, wrap ourselves," "remain where conditions," "world brain," and "such is the vision": J. B. Calhoun, "The Universal City of Ideas," in *The Conference Report, The Exploding Cities*, ed. R. Richter (London: *The Sunday Times* and UN Population Conference, 1974), 301–6.
2. "to come out into the quiet," "site and solitude," "How we resolve this crisis," "the starkness of winter," and "adequate helicopter": Calhoun to Richard Nixon, January 3, 1974, box 4, folder 16, John B. Calhoun Papers, NLM.
3. "I began to retrench" and "deep depression": J. Calhoun, "Looking Backward from 'The Beautiful Ones,'" in *Discovery Processes in Modern Biology: People and Processes in Biological Discovery*, ed. W. Klemm (New York: Krieger, 1977), 63; "period of reflective thinking," "a total immersion," "no phone calls," and "There is no question": 1974–75 Sabbatical at Home, 1974–1975, box 26, folder 49, John B. Calhoun Papers, NLM; "initiate a frontal attack": "Sabbatical Log," 1974–1975, box 26, folder 49, John B. Calhoun Papers, NLM.
4. "Metascience builds on the strengths," "It differs from basic normal," and "processes and phenomena": Last Cycle of Research Inquiry at NIMH Guided by John B. Calhoun 1975–1982 - A Proposal, 1974, box 26, folder 5, John B. Calhoun Papers, NLM; "There are so many variables,": J. Calhoun, "What Sort of Box?," *Man-Environment Systems* 3 (1973): 22; "those critical times," "contacts with older associates," and "preference for social": Progress Report on Behavioral Development and Dissolution in a Social Setting, July 1, 1975 through June 30, 1976, box 17, folders 15–16, John B. Calhoun Papers, NLM.
5. "to understand [how] individual pathology," Progress Report on the Ultimate Behavioral Pathology, July 1, 1974 through June 30, 1975, box 17, folders 15–16, John B. Calhoun Papers, NLM; "Overcrowding through a succession," "A similar

alteration of nervous," and "ideationally related pathology": Last Cycle of Research Inquiry, 1974, NLM; "animal model of the origin": Looking Forward, from a Late Draft in 1992, 1985/1986/1992, box 138, folder 25, John B. Calhoun Papers, NLM.

6. "counteract[ing] the deleterious": Last Cycle of Research Inquiry, 1974, NLM.
7. "their capacities to perceive" and "exhibit comparable behavioral pathologies": Progress Report on Perceptual and Adaptive Processes in Adjustment to Environmental Change, July 1, 1975 through June 30, 1976, box 17, folders 15–16, John B. Calhoun Papers, NLM; "I propose to make the rats" and "In essence, I propose": Last Cycle of Research Inquiry, 1974, NLM; "network(s) of overlapping," "social network(s) (that) may counteract," and "produces greater sensitivity": Progress Report on Perceptual and Adaptive Processes, July 1, 1975 through June 30, 1976, NLM. For a popular book on social networks in nonhumans, see L. A. Dugatkin, *The Well-Connected Animal: Social Networks and the Wondrous Complexity of Animal Societies* (Chicago: University of Chicago Press, 2024). For one of the few social network studies in nonhumans before Calhoun began his work in Universe 34, see D. Sade, "Some Aspects of Parent-Offspring and Sibling Relations in a Group of Rhesus Monkeys, with a Discussion of Grooming," *American Journal of Physical Anthropology* 23, no. 1 (1965): 1–17; "have much smaller brains than humans": J. B. Calhoun, "A Scientific Quest for a Path to the Future," *Populi* 3 (1976): 19–28, 23.
8. "the institutional response": Calhoun, "Looking Backward," 63–64; "So even if you didn't want": Author interview with Norman Slade, August 9, 2022.
9. "encounter different ideas in conjunction": J. B. Calhoun, ed., *Environment and Population: Problems of Adaptation* (New York: Praeger Publishing Co., 1983), 9; "an amalgam of mind": Calhoun, *Environment and Population*, x; "eliciting individuals with demonstrated": Calhoun, *Environment and Population*, xiii; "*℞evolution: Prescription for Evolution*": box 26, folders 45–47, John B. Calhoun Papers, NLM; "*The Rodent Key to Human Survival*": Looking Forward book, Hopkins Press, 1976–1984, box 138, folder 3, John B. Calhoun Papers, NLM, 55; "*Looking Forward*": Looking Forward book, Hopkins Press, 1976–1984, NLM, 3, 4, 5–10, 10–29; "The Beautiful Ones—Overliving," "Induced Cultural Evolution: Rats to 'Ape,'" "Prosthetic Brain R & D," and "The fate of the architect" (Goethe, *Elective Affinities*, bk. II, chap. 3, as cited by Calhoun): Outline of *℞evolution*, Prescription for Evolution, 1974, box 26, folder 47, John B. Calhoun Papers, NLM; "promise[d] to be a more": "Sabbatical Log," 1974–1975, NLM; J. B. Calhoun, "The Role of Brain Prostheses and Organizational Synergy in Information Metabolism," *Psychopharmacology Bulletin* 10 (1974): 28–29; "prosthetic social brains": J. B. Calhoun, "Evolutionary Perspective on Environmental Crisis," *Fields within Fields* 13 (1974): 18–30.
10. E. O. Wilson, *Sociobiology: The New Synthesis* (Cambridge, MA: Harvard University Press, 1975); "Biological determinist arguments": Richard Lewontin, in a

position paper he sent to the press, as quoted in B. Rensberger, "The Basic Elements of the Arguments Are Not New," *New York Times*, November 9, 1975. The other *New York Times* articles on *Sociobiology* were: B. Rensberger, "Sociobiology: Updating Darwin on Behavior," *New York Times*, May 28, 1975; J. Pfeiffer, "Sociobiology," *New York Times*, July 27, 1975; and E. O. Wilson, "Human Decency is Animal," *New York Times*, October 12, 1975; "John B. Calhoun's famous": Wilson, *Sociobiology*, 84; "bizarre effects were observed": Wilson, *Sociobiology*, 255; For more on the debates surrounding *Sociobiology*, see U. Segerstrale, *Defenders of the Truth: The Battle for Science in the Sociobiology Debate and Beyond* (New York: Oxford University Press, 2004).

CHAPTER 15

1. All of the following are from B. Greenbie, *Design for Diversity: Planning for Natural Man in the Neo-Technic Environment* (Amsterdam: Elsevier, 1976): "devised a number of environments" (p. 35), "A Multi-Celled Habitat for Man?" (p. 37), and "The fact that" (p. 41).
2. H. Scarupa, "Of Mice and Men and Escaping the Ultimate Pathology," *Sun Magazine*, June 13, 1976; J. N. Wilford, "Rats on Atoll Seared by Nuclear Tests Are Providing Valuable Data for Scientists," *New York Times*, April 18, 1977; "Early in his career [Calhoun] skipped": W. Klemm, ed., *Discovery Processes in Modern Biology: People and Processes in Biological Discovery* (New York: Krieger, 1977), editor's foreword. Other papers Calhoun wrote around this time include J. B. Calhoun, "Crowding and Social Velocity," in *New Dimensions in Psychiatry: A World View*, vol. 2, eds. S. Arieti and G. Chrzanowski (New York: John Wiley and Sons, 1977), 20–44; J. B. Calhoun and D. R. Ahuja, "Population and Environment: An Evolutionary Perspective to Development," in *World Population and Development: Challenges and Prospects*, ed. P. M. Hauser (Syracuse, NY: Syracuse University Press, 1979), 80–98; and J. B. Calhoun, "Biological Basis of the Family," in *Georgetown Family Symposia*, vol. III (1975–1976), *A Collection of Selected Papers*, ed. R. R. Sagar (Washington, DC: Georgetown University Family Center, 1978), 52–67; J. B. Calhoun, "A Scientific Quest for a Path to the Future," *Populi* 3 (1976): 19–28.
3. "It's cold [and] you walk into this," "To this day," "did nothing," and "the idea of the colony as a brain": Author interview with Kathy Kerr, July 13, 2022.
4. "It now looks as though": *Crowding and Social Dynamics*, Progress Report, July 1, 1977 through September 30, 1978, Annual Reports, 1976–1983, box 17, folders 38–41, John B. Calhoun Papers, NLM; "signs of behavioral disintegration," "entered a phase of temporary," and "evidence from inquiries": Annual Reports Summary, Unit for Research on Behavioral Systems, 1979, Annual Reports, 1978–1984, box 18, folders 19–26, John B. Calhoun Papers, NLM; "extreme [social] withdrawal": Progress Report, Group Organization in Rodents, October 1, 1980, through September 30, 1981, box 17, folders 15–16, John B. Calhoun Papers, NLM.

5. "an unusual 'brainstorming' session," "to discuss a new project," "will give the rats," and "expressed the view": Calhoun discusses new project: "Overcoming Effect of Overcrowding," *ADAMHA News*, August 6, 1976; "given the advantage" and "We're reducing the whole evolution": "Rats Given 'Culture' as a Defense against Overcrowding," *ADAMHA News*, September 22, 1978; "[new] modes," "This leads to a generalization," and "Humans are exposed": Progress Report, Conceptual Adaptation and Evolution, October 1, 1979, to September 30, 1980, box 17, folders 15–16, John B. Calhoun Papers, NLM.
6. "Calhoun's experiments are summarized": D. Trueman, "The Effects of Population Density, Room Size and Group Size on Human Intellectual Task Performance and Emotional Reactivity" (PhD diss., Hofstra University, 1975), 17; T. S. Whatson, "The Social Behaviour of Adult Rats Undernourished in Early Life" (PhD diss., University of Manchester, 1975); J. Brewster, "Transport of Young in the Norway Rat" (PhD diss., McMaster University, 1978); R. Cramer, "Some Effects of School Building Renovation on Pupil Attitudes and Behavior in Selected Junior High Schools" (PhD diss., University of Georgia, 1976); "Calhoun observed several," "pansexual males," "probers," "have flawless coats," and "in a state of physical": L. Drickamer and S. Vessey, *Animal Behavior: Concepts, Processes, and Methods* (Boston: W. Grant, 1981), 427–28; "medical researcher," "mice in an earlier study," "competition became intense," and "who learned little": L. Davidoff, *Introduction to Psychology*, 2nd ed. (New York: McGraw-Hill, 1980), 373; John Slater to Calhoun, April 25, 1979, box 15, folder 33, John B. Calhoun Papers, NLM. Slater wrote Calhoun for photos of Universes that he wanted to include in "a college textbook on general zoology under contract with Addison Wesley," but this book appears not to have been published; "deeply impressed": Jun-Ichiro Takeda to Hoffmann, n.d. [but, based on Hoffman's reply, likely sometime in September 1980], box 20, folder 22, John B. Calhoun Papers, NLM. Takeda initially made contact with Calhoun through Dr. Hans Hoffmann at the NIH; "I have worked with," and "animals do not follow script": Calhoun to Jun-Ichiro Takeda, October 9, 1980, Media Requests, 1980, box 15, folder 42, John B. Calhoun Papers, NLM; "Calhoun Mice to 'Star' in Japanese Films," *ADAMHA News*, December 12, 1980; "One of my colleagues remarked": Media, 1981, Calhoun to Joanna Meyer, March 16, 1981, box 15, folders 59–60, John B. Calhoun Papers, NLM. Also see Media, 1981, Calhoun to Joanna Meyer, March 19, 1981, and Calhoun to George Zimmer Meyer, March 16, 1981, box 15, folders 59–60, John B. Calhoun Papers, NLM; "ecopsychologist . . . [and] leading theorist on crowding," "the study of the consequences," "certain features of," "utopia-turned-hell," "environments which enhance," "Should we look at the components of our cities," and "We are on the leading edge": Smithsonian Associates Lecture: Creating the City, 1980, box 15, folder 41, John B. Calhoun Papers, NLM.
7. "a safety factor": Annual Research on Behavioral Systems, Laboratory of Brain Evolution and Behavior, Intramural Research Program, National Institute of

Mental Health, 1980, Annual Reports, 1978–1984, box 18, folders 19–26, John B. Calhoun Papers, NLM; "developed a slave producing": Summary Documents Reviewing Plans and Course of His Lost Cycle of Research, 1975–1980, box 18, folder 35, John B. Calhoun Papers, NLM.

8. "To the extent that this expectation" and "the negentropy level of this part": Progress Report on Conceptual Prosthesis of the Brain, October 1, 1979, through September 30, 1980, box 70, folder 36, John B. Calhoun Papers, NLM; "'Tradition,' in the biological brain": Annual Reports Summary, Unit for Research on Behavioral Systems, 1979, Annual Reports, 1978–1984, box 18, folders 19–26, John B. Calhoun Papers, NLM; "Toward a Negentropic Model": Current Concepts in Pediatrics Lecture, 1981, box 15, folder 57, John B. Calhoun Papers, NLM.
9. "innovative research on environment": Administrative Award for Meritorious Achievement, 1981, box 135, folder 76, John B. Calhoun Papers, NLM; "The NIMH is drugs": Hitchhiker's Guide to Three Worlds, Master, 1986, box 138, folder 36, John B. Calhoun Papers, NLM. Elsewhere, Calhoun wrote "Mental Health is Drugs. Period!" Calhoun to Dominique DeMenil, April 15, 1986, box 1, folder 14, John B. Calhoun Papers, NLM.

   Calhoun also ran a second spin-off experiment on cooperation and population dynamics. During past studies, when no recordings of vocal communications were made, he and his team had observed anecdotal behavior suggesting that vocalizations decreased as conditions became more crowded. To determine whether the coordination inherent with cooperative behaviors in Universe 34B might favor rats who developed more complex ultrasonic communication to solve the problems they were posed with, Calhoun (working with his colleagues James Hill and others) built two new universes similar to 34A and 34B but with ultrasonic microphones placed at strategic locations.

CHAPTER 16

1. "in addition to his renowned research," "a utopian paradise," "who never involved themselves," "Though they looked intelligent," and "Though these studies": *Animal Populations: Nature's Checks and Balances* (Chicago: *Encyclopaedia Britannica* Educational Corporation, 1983), 22 min.; "Discuss the stages": Promotional Material for *Animal Populations: Nature's Checks and Balances*, *Encyclopaedia Britannica* Educational Corporation, 1983, box 15, folder 74, John B. Calhoun Papers, NLM.
2. "biological brain reached," "the network of interpersonal," "represented the last two," "concerned with thought," "[The] next phase shift in evolution," "During the coming era," and "Machines will more and more represent": J. Calhoun, "The Transitional Phase in Knowledge" (lecture, American Association for the Advancement of Science, Washington, DC, 1982); "In the customary sense," "experiment, based on ideas," "about an important," and "a belief of mine": J. B.

Calhoun, ed., *Environment and Population: Problems of Adaptation* (New York: Praeger Publishing Co., 1983), ix; "so rich that my mind" and "to identify generically": Calhoun, *Environment and Population*, x; "one could then write" and "Extreme care needs": Calhoun, *Environment and Population*, xii; S. Milgram, "The Small-World Problem," *Psychology Today* 2 (1967): 60–67.

3. "Such rephrasing of findings," "in between snatches," and "all roles, all tradition": J. B. Calhoun, "Musings on Passage through the Eye of the Needle," 1983, collection of Catherine Calhoun; "The words relaxed": Progress Report, the Influence of Environmental Setting on Behavior and Population Dynamics, October 1, 1981 through September 30, 1982, box 17, folder 21, John B. Calhoun Papers, NLM; "Engaging in cooperative" and "The process of acquiring": Progress Report *Cooperation Induced Modification of Behavior of Rats*, October 1, 1982 through September 30,1983, box 18, folders 3–4, John B. Calhoun Papers, NLM.
4. "From inception to conclusion": Performance Evaluation of Calhoun's Retirement, Evaluation of Calhoun's Work, July 26, 1983, box 17, folder 21, John B. Calhoun Papers, NLM; "Administration of the Department": Performance Evaluation of Calhoun's Retirement, Evaluation of Calhoun's Work, October 1, 1981, box 17, folder 21, John B. Calhoun Papers, NLM; "the nature of [Calhoun's] investigation," "relevance of such studies," and "[It] provides remedial recommendations": Performance Evaluation of Calhoun's Retirement, April 26, 1983, box 17, folder 21, John B. Calhoun Papers, NLM.
5. "the Hopkins Press would like": Assembly of Documents Describing Events Leading to Calhoun's Retirement under Protest, 1986, box 1, folder 15, John B. Calhoun Papers, NLM; "background material": Looking Forward book, Hopkins Press, 1976–1984, box 138, folder 3, John B. Calhoun Papers, NLM; "apparent fixed decision," "exploring how the," "would not allow," "could put up with," and "current uncertainties": Hitchhiker's Guide to Three Worlds,1986, box 138, folder 36, John B. Calhoun Papers, NLM; "discuss briefly with" and "It grieves me": Calhoun to Donald Ian Macdonald, March 31, 1986, Calhoun Correspondence, 1984–1986, box 1, folder 14, John B. Calhoun Papers, NLM; "My broader view of what humanity": Calhoun to John Herbers, September 3, 1986, Calhoun Correspondence, 1984–1986, box 1, folder 14, John B. Calhoun Papers, NLM.

CHAPTER 17

1. "described a phenomenon in which": H. J. Fountain, "J. B. Calhoun, 78, Researcher on Effects of Overpopulation," *New York Times*, September 29, 1995; "grim forecasts for the future": B. Barnes, "Scientist John Calhoun Dies," *Washington Post*, September 30, 1995; "Gather round my ratties," "Some behaviorists made," "They played," "Rat turned against rat," and "Sickening, isn't it?": A. Grant, J. Balent, and B. Smith, "The Secret of the Universe, Part Two," *Catwoman* 2, no. 26, November 1995.

2. J. C. Adams and E. Ramsden, "Rat Cities and Beehive Worlds: Density and Design in the Modern City," *Comparative Studies in Society and History* 53, no. 4 (2011): 722–56; E. Ramsden, "From Rodent Utopia to Urban Hell: Population, Pathology, and the Crowded Rats of NIMH," *Isis* 102, no. 4 (2011): 659–88; E. Ramsden, "Travelling Facts about Crowded Rats: Rodent Experimentation and the Human Sciences," in *How Well Do Facts Travel?*, eds. M. Morgan and P. Howlett (Cambridge: Cambridge University Press, 2011), 223–51; E. Ramsden, "Rats, Stress and the Built Environment," *History of the Human Sciences* 25, no. 5 (2012): 123–47; and E. Ramsden and J. C. Adams, "Escaping the Laboratory: The Rodent Experiments of John B. Calhoun and Their Cultural Influence," *Journal of Social History* 42, no. 3 (2009): 761–92; "Calhoun's research remains": W. Wiles, "The Behavioral Sink: The Mouse Universes of John B. Calhoun," *Cabinet Magazine* 42 (2011), http://www.cabinetmagazine.org/issues/42/wiles.php; *Critical Mass*, directed by Mike Freedman (Surrey, England: Journeyman Pictures, 2012); "spared from violence and death" and "Can we escape": E. Inglis-Arkell, "How Mice Turned Their Private Paradise into a Terrifying Dystopia," *Gizmodo*, February 24, 2015, www.gizmodo.com/how-rats-turned-their-private-paradise-into-a-terrifyin-1687584457; "extrapolated . . . to human concerns": C. Giaimo, "The Doomed Mouse Utopia That Inspired the 'Rats of NIMH': Dr. John Bumpass Calhoun Spent the '60s and '70s Playing God to Thousands of Rodents," *Atlas Obscura*, September 14, 2016, www.atlasobscura.com/articles/the-doomed-mouse-utopia-that-inspired-the-rats-of-nimh; "Rats and mice" and "Just as Calhoun had": F. Kunkle, "The Researcher Who Loved Rats and Fueled Our Doomsday Fears," *Washington Post*, June 19, 2017.

EPILOGUE

1. "who achieved the recognition": F. Kunkle, "The Researcher Who Loved Rats and Fueled Our Doomsday Fears.," *Washington Post*, June 19, 2017; L. A. Dugatkin, *Principles of Animal Behavior*, 5th ed. (Chicago: University of Chicago Press, forthcoming).
2. So complex are the social dynamics in these barn mice that König and her colleague Patricia Lopes began looking at social networks and what is known as gene expression—that is, what genes are turning on and turning off (in this case, in the brains of mice in the barn). In males, across all brain regions, only three genes showed different expression levels as a function of the number of network ties a male had. But in females, gene-expression levels differed rather dramatically between well-connected mice and their less-connected counterparts in the barn: 180 genes, including some known to be important in learning, showed differences in gene-expression levels. For more on house mouse networks, see B. König, A. K. Lindholm, P. C. Lopes, A. Dobay, S. Steinert, and F. J.-U. Buschmann, "A System for Automatic Recording of Social Behavior in a

Free-Living Wild House Mouse Population," *Animal Biotelemetry* 3 (2015), https://doi.org/10.1186/s40317-015-0069-0; P. C. Lopes, P. Block, and B. König, "Infection-Induced Behavioural Changes Reduce Connectivity and the Potential for Disease Spread in Wild Mice Contact Networks," *Scientific Reports* 6 (2016), https://doi.org/10.1038/srep31790; J. C. Evans, J. I. Liechti, B. Boatman, and B. König, "A Natural Catastrophic Turnover Event: Individual Sociality Matters despite Community Resilience in Wild House Mice," *Proceedings of the Royal Society B: Biological Sciences* 287 (2020), https://doi.org/10.1098/rspb.2019.2880; P. C. Lopes, E. H. D. Carlitz, M. Kindel, and B. König, "Immune-Endocrine Links to Gregariousness in Wild House Mice," *Frontiers in Behavioral Neuroscience* 14 (2020), https://doi.org/10.3389/fnbeh.2020.00010; and J. C. Evans, A. K. Lindholm, and B. König, "Family Dynamics Reveal That Female House Mice Preferentially Breed in Their Maternal Community," *Behavioral Ecology* 33, no. 1 (2022): 222–32; "He anticipated": Author interview with Steve Suomi, July 26, 2022; "I ended up using": Author interview with Neil Greenberg, August 12, 2022.

INDEX

Page numbers in italics refer to figures.